AF795607

E2RA5

Chlorine

Edited by Peter Schmittinger

WILEY-VCH

Chlorine

Principles and Industrial Practice

Edited by Peter Schmittinger

WILEY-VCH

Weinheim · New York · Chichester · Brisbane · Singapore · Toronto

NC-IB

(E2RA5)

Dr. Peter Schmittinger
Wallbergstr. 2
82008 Unterhaching

This book was carefully produced. Nevertheless, editor, authors and publisher do not warrant the information contained therein to be free of errors. Readers are advised to keep in mind that statements, data, illustrations, procedural details or other items may inadvertently be inaccurate.

First Edition 2000

Cover picture: Chlorine tree (courtesy of Euro Chlor)

UML WI HDRAWN

Library of Congress Card No.: Applied for.

British Library Cataloguing-in-Publication Data: A catalogue record for this book is available from the British Library.

Die Deutsche Bibliothek – CIP Cataloguing-in-Publication-Data
A catalogue record for this publication is available from Die Deutsche Bibliothek

© WILEY-VCH Verlag GmbH, D-69469 Weinheim (Federal Republic of Germany), 2000

Printed on acid-free and chlorine-free paper.

All rights reserved (including those of translation in other languages). No part of this book may be reproduced in any form – by photoprinting, microfilm, or any other means – nor transmitted or translated into machine language without written permission from the publishers. Registered names, trademarks, etc. used in this book, even when not specifically marked as such, are not to be considered unprotected by law.

Composition and Printing: Rombach GmbH, D-79115 Freiburg
Bookbinding: Wilhelm Osswald & Co., D-67433 Neustadt (Weinstraße)

Printed in the Federal Republic of Germany

661.113
SCH

JOHN RYLANDS
UNIVERSITY
LIBRARY OF
MANCHESTER

Preface

Chlorine is one of the most important basic products of the chemical industry since a large number of chemicals require chlorine for their production. The total worldwide production of chlorine is currently about 45 million tonnes per year and consumes around 1.3×10^{11} kWh of electrical energy. The production and use of chlorine supports more than 3 million jobs in the United States and western Europe, about 60 % of chemical industry turnover depending on chlorine in developed countries.

Chlorine has major uses in water treatment and as a disinfectant and is heavily used in plastics manufacture, pharmaceuticals and crop protection. However, the public conception of chlorine is largely based on its "poison gas" reputation and its beneficial qualities go unappreciated to a great extent except within the industry itself and by chemists.

This book has been written at the suggestion of WILEY-VCH and is based on the chlorine section of *Ullmann's Encyclopedia of Industrial Chemistry.* The properties, manufacturing processes, uses and handling of chlorine are described in detail and current issues involving the environment, health and toxicology, and economics are dealt with comprehensively. In order to ensure the most up-to-date views and information, each chapter has been written by an acknowledged expert in the field. The many tables and diagrams, along with a full index, make the book suitable for use as a reference while the useful bibliography allows access to the original literature.

The book is intended for chemical technologists in all industries who are involved in the production, use and environmental effects of chlorine. It will also be valuable in universities.

The editor is very grateful to the authors for their excellent cooperation, to Degussa-Hüls AG, the Chlorine Institute, Eurochlor and the Verband der Chemischen Industrie for providing information and literature, and to Ivan Davies for critically reviewing the text.

December 1999 Peter Schmittinger

List of Contributors

Dr. Rüdiger Bartsch
GSF-Forschungszentrum
Institut für Toxikologie
Postfach 1129
85758 Neuherberg
Germany
Chapter 16

Calvert L. Curlin
CURLIN CHLOR-ALKALI
CONSULTANTS
1186 Foxfire Drive
Painesville, Ohio 44077-5238
USA
Chapters 6 and 9 (in part)

Thomas F. Florkiewicz
ELTECH Systems Corp.
100 Seventh Avenue, Suite 300
Chardon, OH 44024-1095
USA
Chapters 6 and 9 (in part)

Dr. Benno Lüke
Krupp Uhde GmbH
Friedrich-Uhde-Str.
44141 Dortmund
Germany
Chapters 7 and 9 (in part)

Dr. Hans-Rudolf Minz
Hans-Sachs-Str. 14
41542 Dormagen
Germany
Chapter 9 (in part)

Thomas Navin
ELTECH Systems Corp.
100 Seventh Avenue, Suite 300
Chardon, OH 44024-1095
USA
Section 8.2

Dr. Robert Scannell
De Nora Deutschland GmbH
Postfach 1553
63405 Hanau
Germany
Section 8.1

Dr. Peter Schmittinger
Wallbergstr. 2
82008 Unterhaching
Germany
Chapters 1–5, 9 (in part), 10, 11, 12 (in part) and 13–15, 17

Dr.-Ing. Erich Zelfel
Infraserv GmbH & Co.
Knapsack KG
Bereich Technik
Industriestr.
50354 Hürth
Germany
Chapter 12

Contents

1. Introduction

Although C. W. SCHEELE reported the formation of chlorine gas from the reaction of manganese dioxide with hydrochloric acid in 1774, he did not recognize the gas as an element [37]. H. DAVY is usually accepted as the discoverer (1808), and he named the gas *chlorine* from the Greek κλῶϱοσ (chloros), meaning greenish yellow. Chlorine for bleaching textiles was first produced from manganese dioxide and hydrochloric acid by a process developed by WELDON, the yield of chlorine being 35 % of the theoretical value. In 1866, DEACON developed a process based on the oxidation of hydrogen chloride gas by atmospheric oxygen in the presence of a copper salt, $CuCl_2$, as the catalyst and obtained yields up to 65 % of the theoretical value.

In 1800, CRUICKSHANK was the first to prepare chlorine electrochemically [38]; however, the process was of little significance until the development of a suitable generator by SIEMENS and of synthetic graphite for anodes by ACHESON and CASTNER in 1892. These two developments made possible the electrolytic production of chlorine, the *chlor-alkali process,* on an industrial scale. About the same time, both the *diaphragm cell* process (1885) and the *mercury cell* process (1892) were introduced. The *membrane cell* process was developed much more recently (1970). Currently, more than 95 % of world chlorine production is obtained by the chlor-alkali process. Since 1970 graphite anodes have been superseded by *activated titanium anodes* in the diaphragm and mercury cell processes. The newer membrane cell process uses only activated titanium anodes.

Other electrochemical processes in which chlorine is produced include the electrolysis of hydrochloric acid and the electrolysis of molten alkali metal and alkaline earth metal chlorides, in which the chlorine is a byproduct. Purely chemical methods of chlorine production are currently insignificant.

Since 1975, the membrane cell process has been developed to a high degree of sophistication. It has ecological advantages over the mercury and diaphragm processes and has become the most economically advantageous process. The membrane cell process has become widely accepted, and all new plants are using this technology. By 2000 more than 30 % of the chlorine worldwide will be produced in membrane cells.

World capacity for chlorine exceeds 45×10^6 t/a. With an annual energy consumption of about 1.5×10^{11} kW h, the chlor-alkali process is one of the largest industrial consumers of electrical energy. The chlorine worldwide production of a country is an indicator of the state of development of its chemical industry.

Occurrence and Formation. Chlorine is the 11th most abundant element in the lithosphere. Because it is highly reactive, it is rarely found in the free state and then mainly in volcanic gases. It exists mainly in the form of chlorides, as in sea water, which contains an average of 2.9 wt% sodium chloride and 0.3 wt% magnesium chloride. In salt deposits formed by evaporation of seas, there are large quantities of rock salt (NaCl) and sylvite (KCl), together with bischofite ($MgCl_2 \cdot 6\,H_2O$), carnallite ($KCl \cdot MgCl_2 \cdot 6\,H_2O$), tachhydrite ($CaCl_2 \cdot 2\,MgCl_2 \cdot 12\,H_2O$), kainite ($KCl \cdot MgSO_4 \cdot$

$3\,H_2O$), and others. Occasionally there are also heavy metal chlorides, usually in the form of double salts, such as atacamite ($CuCl_2 \cdot 3\,Cu(OH)_2$), and compounds of lead, iron, manganese, mercury, or silver. Chlorates and perchlorates occur to a small extent in Chile saltpeter. Free hydrochloric acid is occasionally found in gases and springs of volcanic origin. Plants and animals always contain chlorine in the form of chlorides or free hydrochloric acid.

Chlorine is formed by oxidation of hydrochloric acid or chlorides by such compounds as manganese dioxide, permanganates, dichromates, chlorates, bleaching powder, nitric acid, or nitrogen oxides. Oxygen, including atmospheric oxygen, acts as an oxidizing agent in the presence of catalysts. Some metal chlorides produce chlorine when heated, for example, gold(III) chloride or platinum chloride.

2. Physical Properties

Chlorine [*7782-50-5*], EINECS no. 231-959-5, exists in all three physical states. At STP it is a greenish-yellow pungent, poisonous gas, which liquefies to a mobile yellow liquid. Solid chlorine forms pale yellow rhombic crystals. The principal properties are given below; more details, including thermodynamic values are given in [40] and in "New Property Tables of Chlorine in SI Units" [41]. There are small differences in the values of some properties in different references.

Atomic number Z	17
Relative atomic mass A_r	35.453
Stable isotopes (abundance)	35 (75.53 %) 37 (24.47 %)
Electronic configuration in the ground state	[Ne] $3s^2 3p^5$
Term symbol in the ground state	$^2P_{3/2}$
Melting point *mp*	172.17 K (– 100.98 °C)
Boiling point *bp*	239.02 K (– 33.97 °C) *
Critical density ϱ_{crit}	573.00 kg/m^3 *
Critical temperature T_{crit} (t_{crit})	416,9 K (143,75 °C)
Critical pressure p_{crit}	7977 kPa *
Density of gas ϱ (0 °C, 101.3 kPa)	3.213 kg/m^3
Density relative to air d	2.48
Enthalpy of fusion ΔH_f	90.33 kJ/kg
Enthalpy of vaporization ΔH_v	288.08 kJ/kg *
Standard electrode potential $E°$	1.359 V
Enthalpy of dissociation ΔH_{diss}	239.44 kJ/mol (2.481 eV)
Electron affinity A	364.25 kJ/mol (3.77 eV)
Enthalpy of hydration ΔH_{hyd} of Cl^-	405.7 kJ/mol
Ionization energies ΔE_i	13.01, 23.80, 39.9, 53.3, 67.8, 96.6, 114.2 eV
EC No.	017-001-00-7

* Values adopted from The Chlorine Institute

The density of chlorine gas at 101.3 kPa is a function of temperature:

t, °C	0	50	100	150
ϱ, kg/m^3	3.213	2.700	2.330	2.051

The density up to 300 °C is higher than that of an ideal gas because of the existence of more complex molecules, for example, Cl_4. In the range 400 – 1450 °C, the density approximates that of an ideal gas, and above 1450 °C thermal dissociation takes place, reaching 50 % at 2250 °C. The density of chlorine gas as a function of temperature and

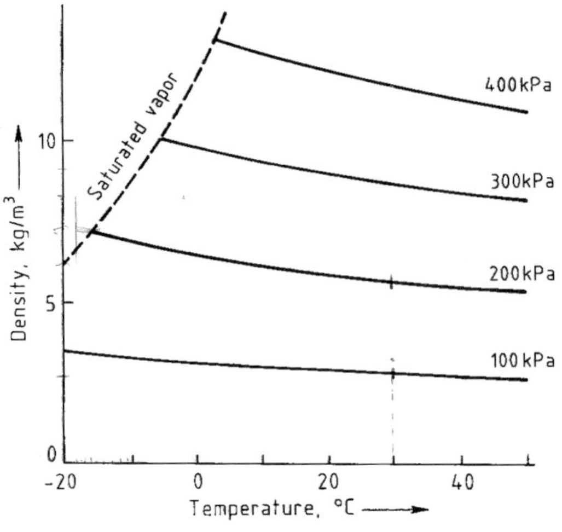

Figure 1. Density of chlorine gas as a function of temperature and pressure

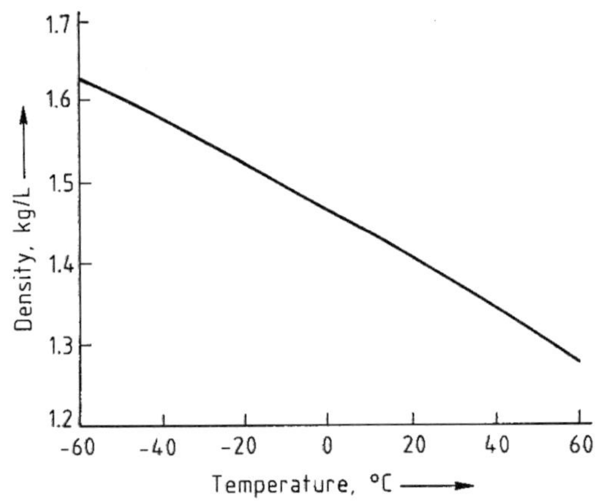

Figure 2. Density of liquid chlorine

pressure is shown in Figure 1. The gas state can be described by the van der Waals equation

$$\left(p+\frac{n^2a}{V^2}\right)(V-nb)=nRT, \text{ with}$$

a = 6.580 L^2 bar mol^{-2}, b = 0.05622 L/mol

The density of liquid chlorine is given in Figure 2. The compressibility of liquid chlorine is the greatest of all the elements. The volume coefficient per MPa at 20 °C over the range 0 – 10 MPa is 0.012 %. The coefficient increases rapidly with temperature: 0.023 % at 35 °C, 0.037 % at 64 °C, and 0.064 % at 91 °C. One liter of liquid chlorine at 0 °C produces 456.8 dm^3 of chlorine gas at STP; 1 kg of liquid produces 311 dm^3 of gas.

The vapor-pressure curve for chlorine is shown in Figure 3.

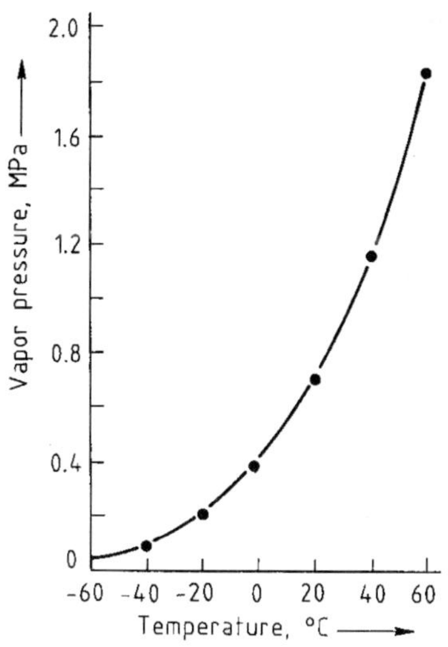

Figure 3. Vapor pressure of liquid chlorine

The vapor pressure can be calculated over the temperature range 172 – 417 K from the Martin – Shin – Kapoor equation [41]:

$$\ln P = A + \frac{B}{T} + C\ln T + DT + \frac{E(F - T)\ln(F - T)}{FT}$$

$A = 62.402508$
$B = -4343.5240$
$C = -7.8661534$
$D = 1.0666308 \times 10^{-2}$
$E = 95.248723$
$F = 424.90$

Thermodynamic information is given in Table 1, from which the data required for working with gaseous and liquid chlorine can be obtained [42]. The Joule – Thomson coefficient is 0.0308 K/kPa at STP.

At STP the specific heats of chlorine are

$c_p = 0.481$ kJ kg^{-1} K^{-1}
$c_v = 0.357$ kJ kg^{-1} K^{-1}
$\kappa = c_p/c_v = 1.347$

The molar heat capacity at constant volume c_v increases with temperature [43]:

t, °C	0	100	200	500	1000
c_v, J/mol	24.9	26.4	28.1	28.9	29.7

Table 1. Properties of liquid and gaseous chlorine [41]. Lower values are quoted in more recent literature [38], [39], especially in the region of the critical points.

Temperature t,	Pressure,	Specific volumes, dm^3/kg		Specific enthalpies, kJ/kg*			Specific entropies, kJ kg^{-1} K^{-1}		
°C	p, bar	liquid	vapor	liquid	vaporization	vapor	liquid	vaporization	vapor
–70	0.1513	0.6042	1563	351.11	306.89	658.00	3.9021	1.5106	5.4127
–60	0.2768	0.6135	894.4	360.69	301.58	662.27	3.9481	1.4147	5.3629
–50	0.4762	0.6233	541.8	370.15	296.29	666.41	3.9917	1.3276	5.3193
–40	0.7772	0.6336	344.9	379.70	290.73	670.43	4.0336	1.2468	5.2804
–30	1.212	0.6445	229.0	389.37	284.95	674.33	4.0737	1.1719	5.2456
–20	1.816	0.6560	157.7	399.21	278.84	678.05	4.1131	1.1015	5.2147
–10	2.628	0.6682	112.1	408.88	272.73	681.61	4.1508	1.0362	5.1870
0	3.689	0.6812	81.89	418.68 **	266.28	684.96	4.1868 **	0.9747	5.1615
10	5.043	0.6951	61.26	428.43	259.67	688.10	4.2215	0.9169	5.1385
20	6.731	0.7100	46.77	438.19	252.80	690.99	4.2546	0.8625	5.1171
30	8.800	0.7261	36.35	447.90	245.72	693.63	4.2873	0.8106	5.0978
40	11.30	0.7435	28.66	457.66	238.31	695.97	4.3183	0.7612	5.0790
50	14.27	0.7627	22.88	467.45	230.53	697.98	4.3480	0.7134	5.0614
60	17.76	0.7837	18.44	477.50	222.07	699.57	4.3781	0.6665	5.0447
70	21.84	0.8073	14.97	487.76	212.90	700.66	4.4074	0.6205	5.0279
80	26.55	0.8339	12.20	498.56	202.60	701.16	4.4376	0.5736	5.0112
90	31.95	0.8646	9.944	510.25	190.79	701.04	4.4665	0.5254	4.9919
100	38.14	0.9010	8.082	523.35	176.85	700.20	4.5004	0.4739	4.9743
110	45.18	0.9456	6.508	537.88	160.14	698.02	4.5372	0.4178	4.9551
120	53.18	1.0039	5.169	554.62	139.59	694.21	4.5787	0.3550	4.9337
130	62.24	1.0890	4.001	575.10	113.30	688.39	4.6277	0.2809	4.9086
140	72.50	1.2624	2.842	603.74	71.18	674.91	4.6934	0.1725	4.8659
144	77.01	1.7631	1.763	642.30	0	642.30	4.7825	0	4.7825

* These values have been calculated in S.I. units according to DIN 1345.

** The enthalpy of liquid chlorine at 0 °C was taken to be H_0 = 418.66 kJ/kg; the entropy of liquid chlorine at 0 °C was taken to be ϱ_0 = 4.1868 kJ kg^{-1} K^{-1}.

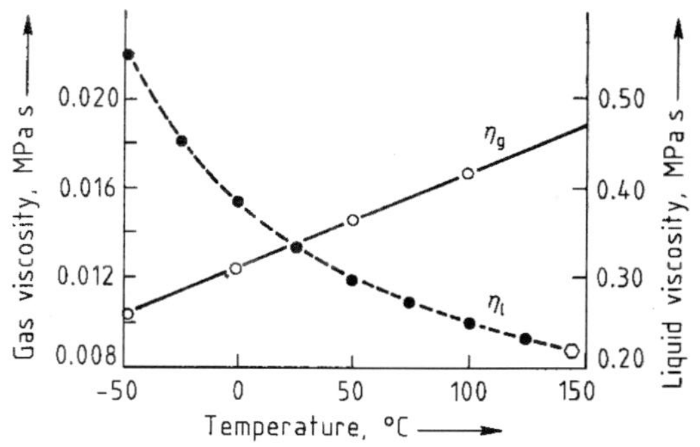

Figure 4. Viscosities of chlorine gas and liquid

The heat capacity of liquid chlorine decreases over the temperature range – 90 °C to 0 °C:

t, °C	– 90	– 70	– 50	– 30	0
c, J kg^{-1} K^{-1}	0.9454	0.9404	0.9341	0.9270	0.9169
c, J mol^{-1} K^{-1}	67.03	66.70	66.23	65.73	65.02

The thermal conductivities of chlorine gas and liquid are almost linear functions of temperature from – 50 °C to 150 °C:

t, °C	– 50	– 25	0	25	50	75	100
λ_g, W m^{-1} K^{-1}x10^2	6.08	7.06	7.95	8.82	9.75	10.63	11.50
λ_l, W m^{-1} K^{-1}	0.17	0.16	0.15	0.135	0.12	0.11	0.09

The viscosities of chlorine gas and liquid are shown in Figure 4 over the same temperature range. The surface tension at the liquid – gas interface falls rapidly with temperature:

t, °C	– 50	– 25	0	25	50
σ, mJ/m^2	29.4	25.2	20.9	16.9	13.4

The specific magnetic susceptibility at 20 °C is -7.4×10^{-9} m^3/kg.

Liquid chlorine has a very low electrical conductivity, the value at – 70 °C being 10^{-16} Ω^{-1} cm^{-1}. The dielectric constant of the liquid for wavelengths greater than 10 m is 2.15 at – 60 °C, 2.03 at – 20 °C, 1.97 at 0 °C, and 1.54 at 142 °C, near the critical temperature.

Chlorine gas can be absorbed in considerable quantities onto activated charcoal and silica gel, and this property can be used to concentrate chlorine from gas mixtures containing it.

Chlorine is soluble in cold water, usually less so in aqueous solutions. In salt solutions, the solubility decreases with salt concentration and temperature. In hydrochloric acid, chlorine is more soluble than in water, and the solubility increases with

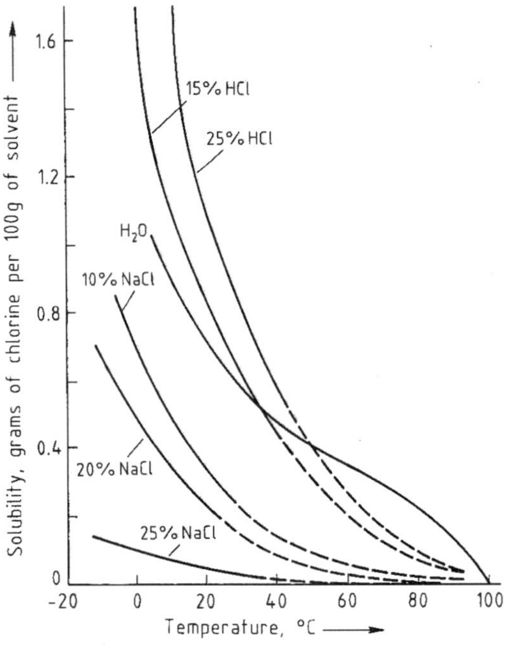

Figure 5. Solubility of chlorine in water, hydrochloric acid (two concentrations), and sodium chloride solutions (three concentrations) All percentages are weight percents.

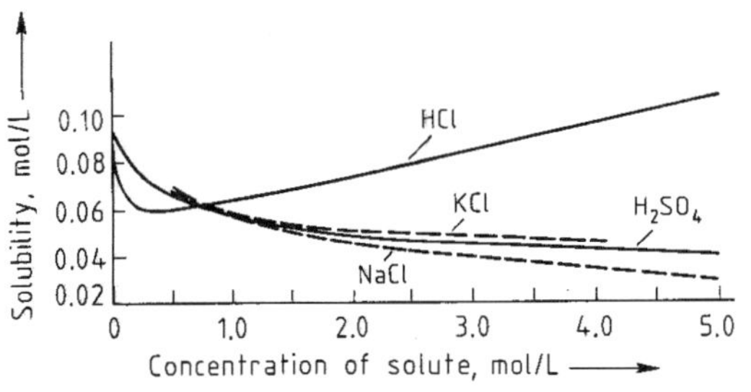

Figure 6. Solubility of chlorine in solutions of KCl, NaCl, H_2SO_4, and HCl at 25 °C

acid concentration (Fig. 5 and Fig. 6). In aqueous solutions, chlorine is partially hydrolyzed, and the solubility depends on the pH of the solution. Below 10 °C chlorine forms hydrates, which can be separated as greenish-yellow crystals. Chlorine hydrate is a clathrate, and there is no definite chlorine:water ratio. The chlorine – water system has a quadruple point at 28.7 °C; the phase diagram has been worked out by KETELAAR [44].

Chlorine is readily soluble in sulfur – chlorine compounds, which can be used as industrial solvents for chlorine. Disulfur dichloride [*10025-67-9*], S_2Cl_2, is converted to sulfur dichloride (SCl_2) and sulfur tetrachloride (SCl_4). Some metallic chlorides and oxide chlorides, such as vanadium oxide chloride, chromyl chloride, titanium tetrachloride, and tin(IV) chloride, are good solvents for chlorine. Many other chlorine-containing compounds dissolve chlorine readily. Examples are phosphoryl chloride, carbon tetrachloride (Fig. 7), tetrachloroethane, pentachloroethane, hexachlorobutadiene

Table 2. Solubility of chlorine in various solvents

Solvent	Temperature, °C	Solubility, wt%
Sulfuryl chloride	0	12.0
Disulfur dichloride	0	58.5
Phosphoryl chloride	0	19.0
Silicon tetrachloride	0	15.6
Titanium tetrachloride	0	11.5
Benzene	10	24.7
Chloroform	10	20.0
Dimethylformamide	0	123 *
Acetic acid, 99.84 wt%	15	11.6 *

* g/100 cm^3

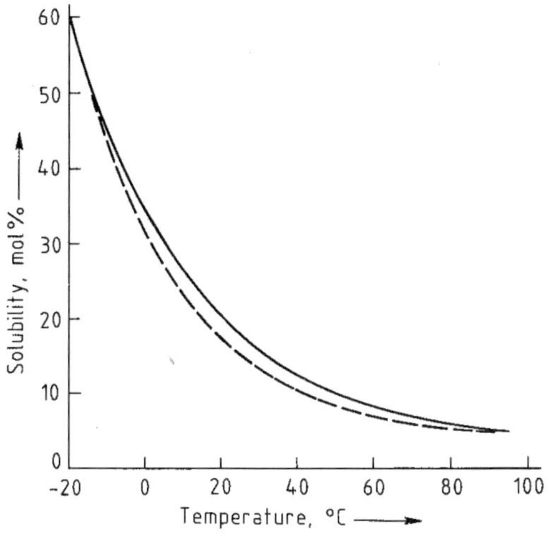

Figure 7. Solubility of chlorine in hexachlorobutadiene (——) and carbon tetrachloride (– – –) at 101 kPa as a function of temperature

(Fig. 7), and chlorobenzene. Chlorine also dissolves in glacial acetic acid, dimethylformamide, and nitrobenzene. The solubility of chlorine in a number of these solvents is given in Table 2.

3. Chemical Properties

Inorganic Compounds. Chlorine, fluorine, bromine, and iodine constitute the halogen group, which has marked nonmetallic properties. The valence of chlorine is determined by the seven electrons in the outer shell. By gaining one electron, the negatively charged chloride ion is formed; the chloride ion has a single negative charge and a complete shell of electrons (the argon structure). By sharing one to seven electrons from the outer shell with other elements, the various chlorine oxidation states can be formed, for example, in the oxides of chlorine, hypochlorites (+ 1), chlorates (+ 5), and perchlorates (+ 7).

The bonds between chlorine and the other halogens are mainly covalent. In the chlorine – fluorine compounds ClF and ClF_3, there is some ionic character to the bond, with chlorine the anion, and in the chlorine – iodine compounds ICl_3 and ICl, there is some ionic character to the bond, with chlorine the cation. Chlorine is very reactive, combining directly with most elements but only indirectly with nitrogen, oxygen, and carbon. Excess chlorine in the presence of ammonia salts forms the very explosive nitrogen trichloride, NCl_3. Hypochlorites react with ammonia to produce the chloramines NH_2Cl and $NHCl_2$. Oxygen and chlorine form several chlorine oxides (→ Chlorine Oxides and Chlorine Oxygen Acids).

Chlorine gas does not react with hydrogen gas [*1333-74-0*] at normal temperatures in the absence of light. In sunlight or artificial light of wavelength ca. 470 nm or at temperatures over 250 °C, the two gases combine explosively to form hydrogen chloride. The explosive limits of mixtures of pure gases lie between ca. 8 vol % H_2 and ca. 14 vol % Cl_2 (the detonation limits). The limits depend on pressure, and the detonation range can be reduced by adding inert gases, such as nitrogen or carbon dioxide (Fig. 8) [45], [46].

Chlorine reacts vigorously with ammonia

$$3\ Cl_2 + 4\ NH_3 \longrightarrow NCl_3 + 3\ NH_4Cl$$

In the presence of the catalyst bromine, chlorine reacts with nitric oxide to give nitrosyl chloride

$$NO + 0.5\ Cl_2 \longrightarrow NOCl$$

Sulfur dioxide and chlorine in the presence of light or an activated carbon catalyst react to form sulfuryl chloride, SO_2Cl_2. Under these conditions carbon monoxide and chlorine react to produce the colorless, highly toxic carbonyl chloride (phosgene), $COCl_2$.

Chlorine reacts with sodium cyanide and sodium thiocyanate to produce cyanogen chloride and thiocyanogen chloride. The reaction of chlorine with sodium thiosulfate [*7772-98-7*] (Antichlor) is used to remove free chlorine from solutions.

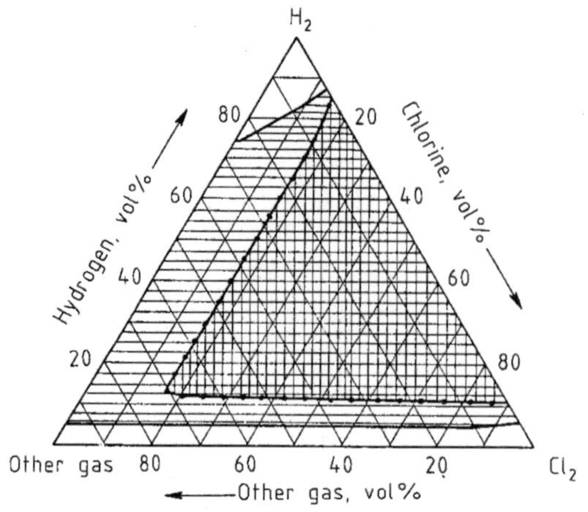

Figure 8. Explosive limits of chlorine–hydrogen–other gas mixture
Horizontally hatched area = Explosive region with residue gas from chlorine liquefaction (O_2, N_2, CO_2)
Checkered area = Explosive region with inert gas (N_2, CO_2)

$$Na_2S_2O_3 + 4\ Cl_2 + 5\ H_2O \longrightarrow 2\ NaHSO_4 + 8\ HCl$$

Chlorine reacts with carbon disulfide to produce carbon tetrachloride and disulfur dichloride.

$$CS_2 + 3\ Cl_2 \longrightarrow CCl_4 + S_2Cl_2$$

The reaction of chlorine with phosphorus produces phosphorus trichloride (PCl_3) and pentachloride (PCl_5). Wet chlorine attacks most metals to form chlorides. Although titanium [*7440-32-6*] is resistant to wet chlorine, it is rapidly attacked by dry chlorine. Tantalum is resistant to both wet and dry chlorine. Most metals are resistant to dry chlorine below 100 °C, but above a specific temperature for each metal, combustion takes place with a flame. This specific temperature, the ignition temperature, also depends on the particle size of the metal so that the following values are only approximate: iron at 140 °C, nickel at 500 °C, copper at 200 °C, and titanium at 20 °C.

Most metal chlorides are soluble in water [3, p. 668], notable exceptions being those of silver (AgCl) and mercury (Hg_2Cl_2). Chlorine liberates bromine and iodine from metallic bromides and iodides, but is itself liberated from metal chlorides by fluorine.

$$0.5\ Cl_2 + KBr \longrightarrow KCl + 0.5\ Br_2$$

Selenium and tellurium react spontaneously with liquid chlorine, whereas sulfur begins to react only at the boiling point. Liquid chlorine reacts vigorously with iodine, red phosphorus, arsenic, antimony, tin, and bismuth. Potassium, sodium, and magnesium are unaffected in liquid chlorine at temperatures below – 80 °C. Aluminum is unattacked until the temperature rises to – 20 °C, when it ignites. Gold is only slowly attacked by liquid chlorine to form the trichloride ($AuCl_3$). Cast iron, wrought iron, carbon steel, phosphor bronze, brass, copper, zinc, and lead are unaffected by dry liquid chlorine, even in the presence of concentrated sulfuric acid.

Organic Compounds. The chlorine-carbon bond is covalent in nature, but the strong electronegativity chlorine (3,2) produces a polar component with a shift of the negative charge in the direction of the chlorine.

$$R_3\overset{\delta+}{C}-\overset{\delta-}{Cl}$$

Chlorine reacts with hydrocarbons either by substitution or by addition. In saturated hydrocarbons, chlorine replaces hydrogen, either completely or partially, to form chlorinated hydrocarbons and hydrogen chloride, e.g. depending on conditions, methane can be chlorinated in stages from methyl chloride (a), to methylene chloride (b), to chloroform (c), to carbon tetrachloride (d):

$$\underset{\text{Methane}}{CH_4} \longrightarrow \underset{(a)}{CH_3Cl} \longrightarrow \underset{(b)}{CH_2Cl_2} \longrightarrow \underset{(c)}{CHCl_3} \longrightarrow \underset{(d)}{CCl_4}$$

In the reaction with unsaturated hydrocarbons chlorine is added to the double or triple bond yielding dichloro- or tetrachloro hydrocarbons, respectively:

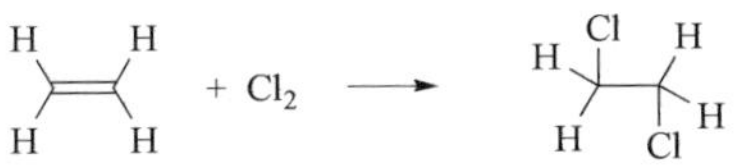

Ethen 1,2-Dichloroethan

In industry the reaction velocity is increased by light (photochlorination), heat, or catalysts.

In aromatic hydrocarbons, both addition and substitution is possible, depending on the conditions (light, temperature, pressure, catalysts).

The reactions of chlorine with toluene demonstrate, in which way the selectivity of the reaction can be directed.

Under the influence of light the hydrogen in the methyl group is substituted by chlorine (radical substitution), forming benzyl chloride (a), benzal chloride (b), and benzotrichloride (c).

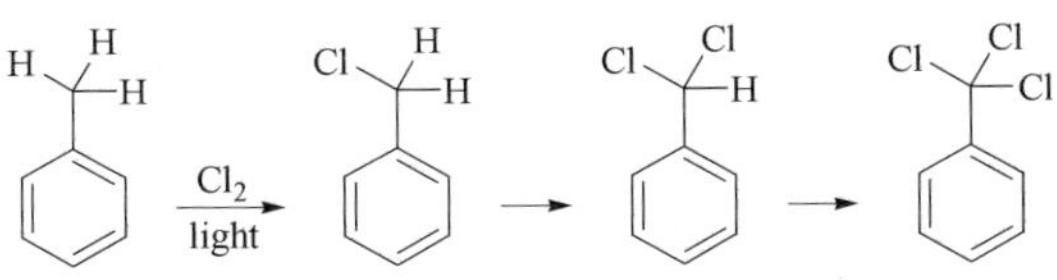

Toluene (a) (b) (c)

In the absence of light, in the presence of a Lewis acid (e.g. $FeCl_3$ or $AlCl_3$) however, the hydrogen in the aromatic system is substituted (electrophilic substitution).

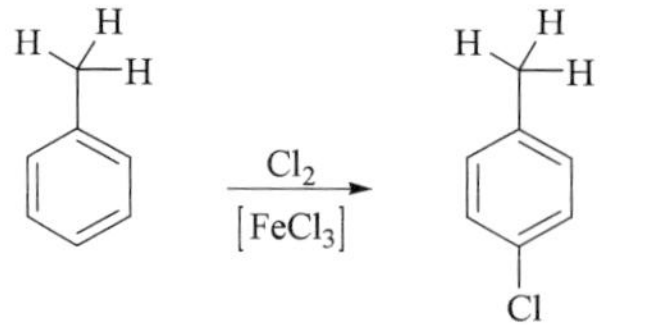

Toluene 4-Chloro-toluene

The reactivity of chlorine with hydrocarbons is high, the energy requirement (pressure, temperature) is comparatively low.

Reactions of Chlorinated Hydrocarbons. The energy of dissociation of the C–Cl bond is relatively low, the manner of dissociation can be heterolytic (ionic) (a), and homolytic (radical) (b).

$$R_3C{-}Cl \longrightarrow R_3C^+ + Cl^- \quad \text{(a)} \qquad\qquad R_3C{-}Cl \longrightarrow R_3C\cdot + Cl\cdot \quad \text{(b)}$$

This fact can be used to replace the chlorine in a chlorinated molecule via a nucleophilic substitution reaction by an other atom or functional group, for example by other halogens like iodine, by alkyl groups, ethers, thioethers, cyano groups, amines etc. (see Fig. 9).

Thus chlorinated hydrocarbons are very useful components for synthesis reactions and as intermediates.

By supplying energy (heat), hydrogen chloride can be eliminated from the chlorinated hydrocarbon. For instance the splitting of HCl from 1,2 dichloroethane forms the vinylchloride monomer.

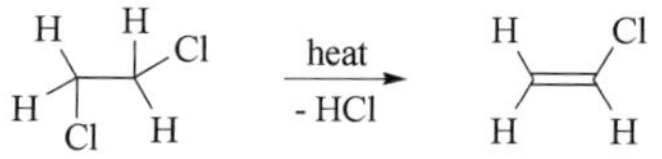

The combination of a chlor-alkali electrolysis (see Chapter 14) with organic chlorination processes offers an elegant and frequently used chance for the synthesis of numerous chlorine containing or chlorine free products. The entire process can be illustrated with the following general scheme (see Fig. 10) [304]:

The electrolysis process converts the electrical energy into chemical energy, which is stored in the chlorine. The subsequent processes then occur with little or no energy consumption.

Summary: The chemistry with chlorine offers the following advantages:

1) Selectivity with high yields of the desired product
2) Reactivity with low consumption of energy
3) Sustainability: the raw material sodium chloride is available in almost an unlimited amount.

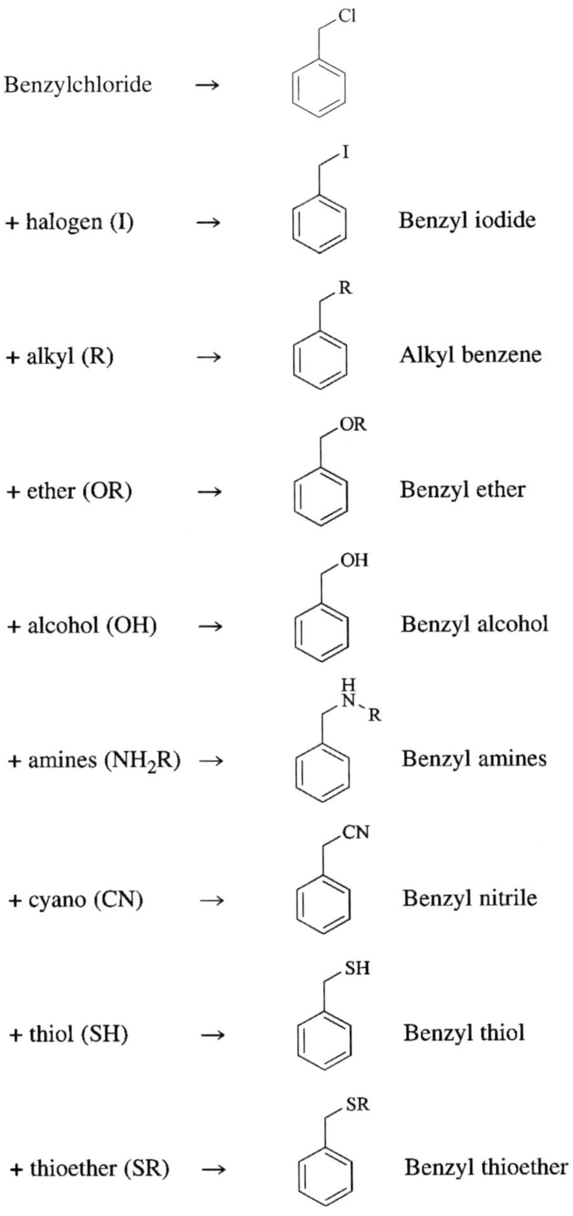

Figure 9. Nucleophilic substitutions in the benzychloride molecule

Chlorine in Nature. Volcanic eruptions are the sources of great masses of hydrogen chloride, they also contain free chlorine gas. Breaking waves and winds over the oceans produce large amounts of seasalt aerosols in the atmosphere, corresponding to ten billion tons of chloride per year. These aerosols are partly deposited on the continents.

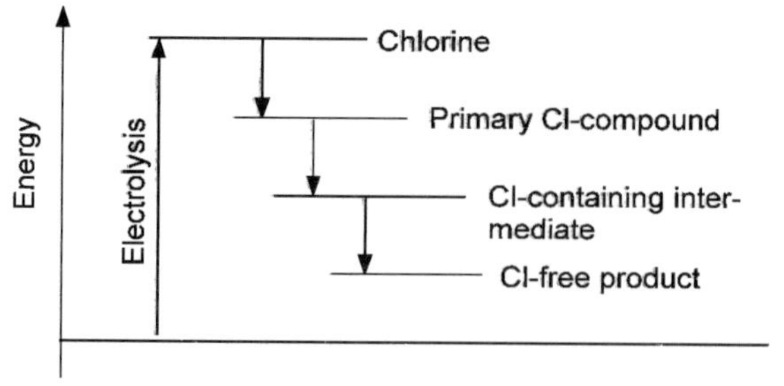

Figure 10. Energy levels in chlorine chemistry (schematic)

Multi-phase photochemical processes in the marine boundary layer convert a portion of the particulate chlorine to volatile products. This process is the major global source of gaseous chlorine in the troposphere. In [305] the planetary chlorine reservoirs (atmosphere, pedosphere and hydrosphere) and the processes transferring chlorine among these reservoirs are discussed in detail.

Despite the wide distribution of halogens, especially of chlorine in nature, the knowledge of the existence of natural occurring chlororganica was scarce until the seventies. This fact led to the assumption, that nature would avoid the use of chlorine in natural substances.

Improved techniques of analysis, better understanding of biological/chemical processes in nature, and systematic research revealed up to 1998 more than 2500 natural halogen compounds, 1800 of them containing chlorine. A further 30 to 40 compounds are detected annually.

In addition to the above mentioned chemical processes, algae, fungi, lichens, bacteria, plants and animals act as terrestrial producers, while in the oceans algae, bacteria, sponges and some fishes synthesize chlororganica.The formation of these substances is catalysed by enzymes like chloro-peroxidases, which are oxidizing the halide ions to form elemental halogens by means of hydrogen peroxide. These elemental halogens are then biologically available.

The spectrum of chlorinated compounds reaches from simple chlorinated C_1-,C_2-, C_3-alkanes to chlorinated terpenes, steroids, fatty acids, alkaloids, heterocycles, polyacetylenes, quinones, phenols to very complex structures like the vancomycin. The structures of some chlororganica are illustrated in Figure 11) [306], [307].

These substances are useful as a deterrent to food competitors, they react bactericidally, fungicidally, or insecticidally, they serve as baits or as growth hormones. The pharmaceutical industry is interested in natural occurring active substances against tumours, in analgesics (there is a frog in Middle America, who is able to synthesize Epibatidine, an analgetic which is 200 times more effective than morphine!), in antibiotics etc.

Despite the continuous formation of great amounts of chlororganica for thousands of years, the relevant enrichment of these substances in nature is not observed. This fact demonstrates, that these substances are decomposed more or less easily, depending on their structure. The decomposition is effected by bacteria, moulds and putrefactive

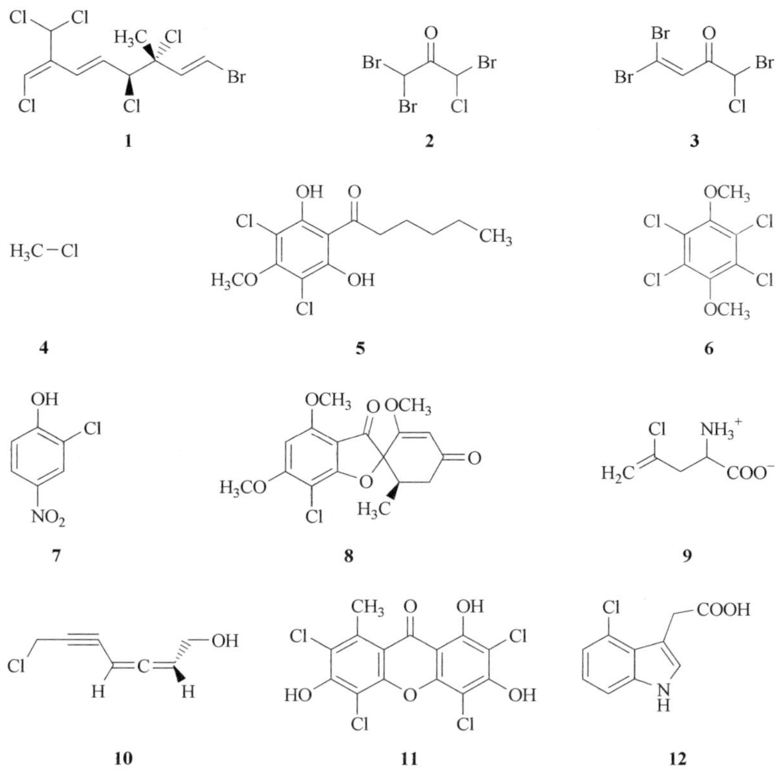

Figure 11. Natural Organochlorines

1) Polyhalogenated monoterpenes from sea hare *Aplysia californica*,
2) 1,1,3-tribromo-3-chloropropane-2-on,
3) 1,4,4-tribromo-1-chloro-butene-2-on in the red algae species *Asparagopsis*,
4) Monochloromethane,
5) Signal substance of the *fungus Dictyostelium discoideum*,
6) Substance in several fungi, e.g. mushrooms,
7) 2-chloro-4-nitrophenol (substance in fungi),
8) Griseofulvin from *Penicillium griseofulvum*,
9) L-2-amino-4-chloro-4-pentene acid,
10) Scorodonin (substance from fungus Marasmius Scorodonius, hinders growth of certain cancer cells),
11) Thiophane acid from lichens,
12) 4-chloroindole-3-acetic acid from peas

fungi, which are able to split the chlorinated hydrocarbons in reductive (anaerobic), oxidative (aerobic) or hydrolytic processes with the formation of hydrogen chloride.

The chloromethane is the compound which is produced in largest masses: the annual production rate is estimated to be more than five million tonnes, 70 % of this mass is created by marine algae and bacteria, the rest by corrosion of dead wood and by burning of biomass.

4. Chlor-Alkali Process

In the chlor-alkali electrolysis process, an aqueous solution of sodium chloride is decomposed electrolytically by direct current, producing chlorine, hydrogen, and sodium hydroxide solution. The overall reaction of the process

$$2\ NaCl + 2\ H_2O \longrightarrow Cl_2 + H_2 + 2\ NaOH$$

takes place in two parts, at the anode and at the cathode. The evolution of chlorine takes place at the anode:

$$2\ Cl^- \longrightarrow 2\ Cl + 2\ e^- \longrightarrow Cl_2 + 2\ e^-$$

There are three basic processes for the electrolytic production of chlorine, the nature of the cathode reaction depending on the specific process. These three processes are (1) the diaphragm cell process (Griesheim cell, 1885), (2) the mercury cell process (Castner – Kellner cell, 1892), and (3) the membrane cell process (1970).

Each process represents a different method of keeping the chlorine produced at the anode separate from the caustic soda and hydrogen produced, directly or indirectly, at the cathode.

These three processes are described in detail in the following three chapters. The basic flow sheets of the three processes are shown in Figures 12 – 14. In all three processes, nearly saturated, purified brine is introduced into the electrolysis cell.

The *hydrogen* produced is cooled as it leaves the decomposer or the cathode compartment and is carried through electrically insulated pipework to a vessel fitted with a water seal (Fig. 15). If a hydrogen – air mixture forms because of a shutdown or breakdown, the seal allows the mixture to escape. A demister ensures that the gas is free of spray, whether water or sodium hydroxide solution. The hydrogen is compressed by Roots-type blowers or reciprocating compressors before it passes through coolers on its way to the consuming plants. At no stage is the pressure allowed to fall below ambient pressure.

Electrolytic hydrogen is very pure, > 99.9 %; however, unwanted traces of oxygen can be removed by reaction with the hydrogen over a platinum catalyst. The hydrogen is used for organic hydrogenation, catalytic reductions, ammonia synthesis and to provide hot flames or protective atmospheres in welding technology, metallurgy, or glass manufacture. It is also used in the manufacture of high-purity hydrogen chloride by combustion with chlorine and as a fuel for heating and drying.

In the *mercury cell process,* sodium amalgam is produced at the cathode. The amalgam is reacted with water in a separate reactor, called the decomposer, to produce hydrogen gas and caustic soda solution.

Because the brine is recirculated, solid salt is required for resaturation. The brine, which must be quite pure, is first dechlorinated and then purified by a straightforward precipitation – filtration process.

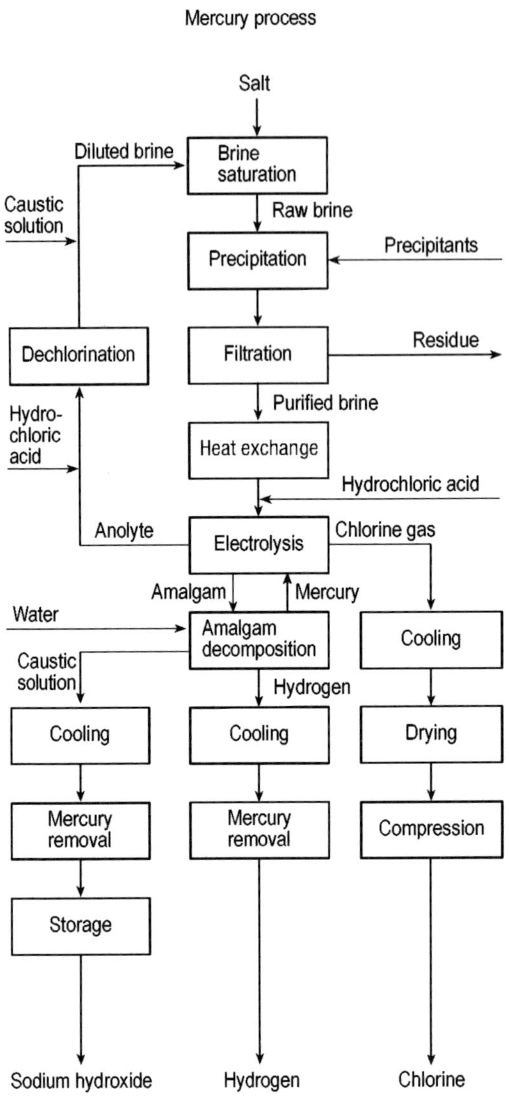

Figure 12. Flow diagram of the chlor-alkali mercury process

The products are extremely pure. The chlorine, along with a little oxygen, generally can be used without further purification. The sodium hydroxide solution contains little chloride and leaves the decomposer with a 50 wt % concentration.

Of the three processes, the mercury process uses the most electric energy; however, no steam is required to concentrate the caustic solution. The use of large quantities of mercury demands measures to prevent environmental contamination. In addition, the hydrogen gas and sodium solution must be freed from mercury. Generally, the operation of the cells is not simple.

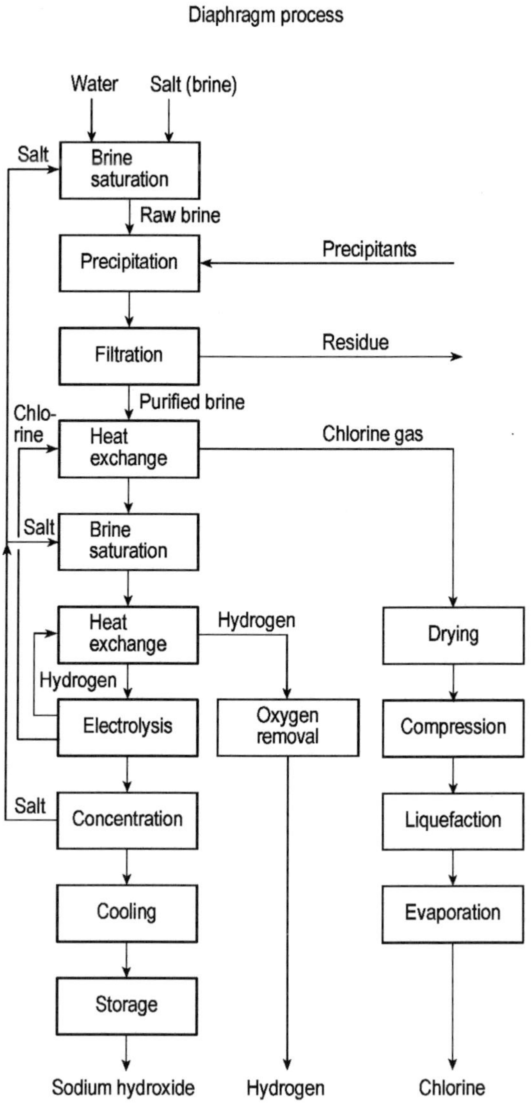

Figure 13. Flow diagram of the chlor-alkali diaphragm process

In the *diaphragm cell process,* the anode area is separated from the cathode area by a permeable, generally asbestos-based diaphragm. The brine is introduced into the anode compartment and flows through the diaphragm into the cathode compartment. Cheaper solution-mined brine can be used; the brine is purified by precipitation – filtration.

A caustic brine leaves the cell, and this brine must be freed from salt in an elaborate evaporative process. Even so, the resultant 50 wt% sodium hydroxide solution contains up to 1 wt% NaCl. The salt separated from the caustic brine can be used to saturate

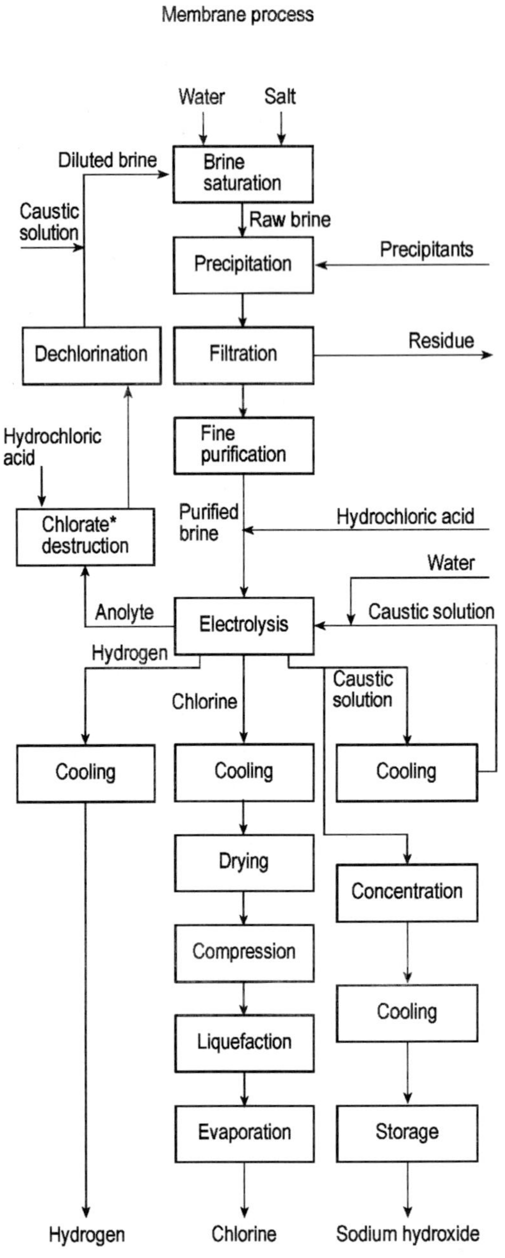

Figure 14. Flow diagram of the chlor-alkali membrane process
(* optional)

dilute brine. The chlorine contains oxygen and must be purified by liquefaction and evaporation.

The consumption of electric energy with the diaphragm cell process is ca. 15 % lower than for the mercury process, but the total energy consumption is higher because of the

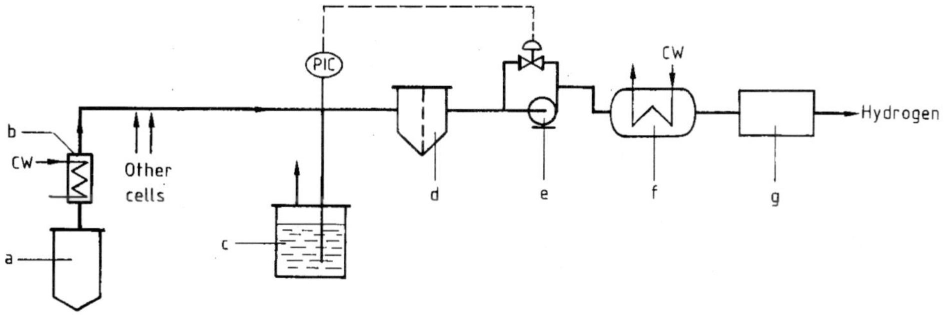

Figure 15. Processing of hydrogen gas from the amalgam decomposer
a) Vertical decomposer; b) Individual cell hydrogen cooler; c) Safety seal; d) Demister; e) Blower; f) Final hydrogen cooler; g) Mercury removal equipment

steam required to concentrate the caustic brine (see Fig. 72). Environmental contamination with asbestos must be avoided. Under constant operating conditions, cell operation is relatively simple.

In the *membrane cell process,* the anode and cathode are separated by a cation-permeable ion-exchange membrane. Only sodium ions and a little water pass through the membrane.

As in the mercury process, the brine is dechlorinated and recirculated, which requires solid salt to resaturate the brine. The life of the expensive membrane depends on the purity of the brine. Therefore, after purification by precipitation – filtration, the brine is also purified with an ion exchanger.

The caustic solution leaves the cell with a concentration of 30 – 36 wt % and must be concentrated. The chloride content of the sodium hydroxide solution is almost as low as that from the mercury process. The chlorine gas contains some oxygen and must be purified by liquefaction and evaporation.

The consumption of electric energy with the membrane cell process is the lowest of the three processes, ca. 25 % less than for the mercury process, and the amount of steam needed for concentration of the caustic is relatively small (see Fig. 72). The energy consumption should be even lower when oxygen-consuming electrodes become common. There are no special environmental problems.

Table 3. Impurities in rock salt and sea salt, wt %

	Rock salt	Sea salt
Insolubles	≤ 2	0.1 – 0.3
Water	≤ 3	2.0 – 6.0
Calcium	0.2 – 0.3	0.1 – 0.3
Magnesium	0.03 – 0.1	0.08 – 0.3
Sulfate	≤ 0.8	0.3 – 1.2
Potassium	≤ 0.04	0.02 – 0.12

4.1. Brine Supply

The brine used in the mercury cell and membrane cell processes is normally saturated with solid salt although there are some installations that use solution-mined brine on a once-through basis. The brine supply for diaphragm cells is always used on a once-through basis, although the salt recovered from caustic soda evaporators may be recycled into the brine supply.

Salt. The basic raw material for the mercury cell and membrane cell processes is usually solid salt. This may be obtained from three sources: rock salt, solar salt, or vacuum-evaporated salt from purifying and evaporating solution-mined brine.

In the United States and Europe, rock salt is most commonly used. The most important impurities are shown in Table 3. The concentrations of these impurities depend on the method of production and on the different grades: crude rock salt, prepared rock salt, and evaporated salt. Solar salt is used in Japan and many other parts of the world, the most important sources being Australia, Mexico, China, Chile, India, and Pakistan. The salt produced by solar evaporation is usually much less pure than rock salt. In a few cases the salt may be obtained from other processes, such as caustic soda evaporation in the diaphragm process.

A new upgrading process (Salex) has been developed by Krebs Swiss [48]. It removes the impurities by selective cracking of the salt crystals and a washing process. Salt losses are minimized, and the purity exceeds 99.95 % NaCl.

Brine Resaturation. In older plants, the open vessels or pits used for storing the salt are also used as resaturators. The depleted brine from the cells is sprayed onto the salt and is saturated, the NaCl concentration reaching 310 – 315 g/L. Modern resaturators are closed vessels, to reduce environmental pollution [49], which could otherwise occur by the emission of a salt spray or mist. The weak brine is fed in at the base of the resaturator, and the saturated brine is drawn off at the top. If the flow rates of the brine and the continously added salt are chosen carefully, the differing dissolution rates of NaCl and $CaSO_4$ result in little calcium sulfate dissolving within the saturator [50]. Organic additives also reduce the dissolution rate of calcium sulfate [51]. The solubility (g per 100 g of H_2O) of NaCl in water does not increase much with temperature (t, °C), whereas the solubility of KCl does:

t	0	20	40	60	80	100
c_{NaCl}	35.6	35.8	36.4	37.0	38.5	39.2
c_{KCl}	28.2	34.4	40.3	45.6	51.0	56.2

Brine Purification. In mercury cells, traces of heavy metals in the brine give rise to dangerous operating conditions (see p. 32), as does the presence of magnesium and to a lesser extent calcium [52]. In membrane cells, divalent ions such as Ca^{2+} or Mg^{2+} are harmful to the membrane. The circulating brine must be rigorously purified to avoid any buildup of these substances to undesirable levels [7]. Calcium is usually precipitated as the carbonate with sodium carbonate; magnesium and iron, as hydroxides with sodium hydroxide; and sulfate, as barium sulfate.

The reagents are usually mixed with weak brine and added to the brine stream at a controlled rate. If solar salt is used, treatment costs may be reduced by prewashing the salt [53]. In order to precipitate calcium at low pH, sodium bicarbonate [54] or phosphoric acid [55] can be added.

The sulfate content can be reduced without the use of expensive barium salts by discharging a part of (purging) the brine [56], by crystallization of $Na_2SO_4 \cdot 10\ H_2O$ on cooling the brine stream [57], by precipitation of the double salt $Na_2SO_4 \cdot CaSO_4$ [58], by an ion-exchange process, or by membrane nanofiltration [59]. Hoechst [60] has a process for recovering barium sulfate of pure pigment quality by precipitation under acid conditions. Chlorate buildup can be avoided by addition of sodium metabisulfite $Na_2S_2O_5$ [61].

After stirring for 1 – 2 h, the precipitated impurities are removed by filtration alone or by sedimentation followed by filtration. Sedimentation is carried out in large circular settling tanks, from which the slurry is removed by mechanical raking equipment, e.g., Clariflocculator, Cyclator, or Dorr thickener. Filtration is carried out with a sand filter, a pressure-leaf filter with filter cloths of chlorine-resistant fabrics, or candle filters automatically cleaned by backflow of brine. The filter is cleaned by water jets, vibrating, or shaking. The separated filter cake is concentrated to 60 – 80 % solids content in rotary drum vacuum filters or centrifuges before disposal. Any soluble material present may be removed from the sludge by washing with water. Barium salts may be recovered by treating with sodium carbonate under pressure [62]. The purified brine should contain ideally $c_{Ca} < 2$ mg/L, $c_{Mg} < 1$ mg/L, and $c_{SO_4} < 5$ g/L.

In the diaphragm process, the removal of sulfate is not always necessary because SO_4^{2-} can be removed from the cell liquor as pure Na_2SO_4 during the concentration process. In the membrane process, the brine must be purified to a much higher degree to avoid the deterioration of the membrane. The Ca^{2+} and Mg^{2+} concentration must be < 0.02 ppm (20 ppb), so a second, fine purification step is required (see Section 7.2.1).

Before the brine enters the electrolysis cells, it should be acidified with hydrochloric acid to pH < 6, which increases the life of the titanium anode coating, gives a purer chlorine product with higher yield, and reduces the formation of hypochlorite and chlorate in the brine.

Brine Dechlorination. In the mercury and membrane processes, the depleted brine leaving the cells must be dechlorinated before resaturation. Further acidification with hydrochloric acid to pH 2 – 2.5 reduces the solubility of chlorine by shifting the equilibrium point of hydrolysis and inhibits the formation of hypochlorite and chlorate. Chlorine discharged in the anolyte tank prior to dechlorination may be fed into the chlorine system. The dissolved chlorine of the brine then is still 400 – 1000 mg/L, depending on pH and temperature. The brine is passed down a packed column or sprayed into a vacuum of 50 – 60 kPa, which reduces the chlorine concentration in the brine to 10 – 30 mg/L. The vacuum is produced by steam jet or liquid-ring vacuum pump. The pure chlorine gas obtained is fed into the chlorine stream.

The water that evaporates from the dechlorinated brine is condensed in a cooler. The condensate, which may be chemically dechlorinated, is returned to the brine circulation system if necessary to maintain the volume of the brine circuit. If necessary, the remaining chlorine content can be further reduced by blowing with compressed air, by a second vacuum treatment, by treatment with activated carbon [63], or by chemical treatment with hydrogen sulfite, thiosulfate, sulfur dioxide, or sodium hydrogensulfide.

Brine Monitoring. The sodium chloride concentration in the brine is determined by density measured by equipment involving radioactive isotopes, vibration techniques, hydrometry, or weighing.

The pH following alkali or acid additions is determined with glass electrodes, and the redox potential following chlorine removal is determined with metal electrodes. Excess OH^- and CO_3^{2-} ions ensure adequate precipitation of dissolved calcium, iron, and magnesium. After filtration, a test sample of 100 mL should require 4 – 6 mL of 0.1 N acid to reach the phenolphthalein end point and a further 0.5 – 1.5 mL to reach the methyl orange end point. Inadequate filtration is detected by turbidimetry in transmitted light or by the Tyndall effect. Calcium and magnesium are determined hourly, and chlorate and sulfate about once per day, all by titration.

4.2. Electricity Supply

Since 1960 the direct current for electrolysis has been provided exclusively by silicon rectifiers. A set of rectifiers can supply up to 450 000 A. Voltages up to 4.0 kV per diode are feasible, but usually for safety, a peak a.c. voltage of 1500 V, corresponding to a d.c. output of 1200 V, is not exceeded. Liquid cooling of the diodes permits a compact design, and self-contained equipment reduces leakage losses. Modern electrolysis cell plants also use continuously variable thyristor converters in place of silicon diodes [64].

Rectification equipment is required to provide steady direct current at a voltage determined by the cell room. The current must remain steady even though the voltage is varied both by the operating condition of the cells and by the number of cells operating. The rectifier equipment usually consists of

transformer capable of variable output voltage with adequate compensation for changing input voltage
silicon rectifiers or thyristors
constant-current control gear
transducers for metering and control
control panels
isolators
cooling equipment
ancillary safety and monitoring equipment

Each set of rectifiers is connected through high-voltage switchgear to the three-phase supply [65]. Smaller units use a 10 – 30 kV supply, but large units can be connected into the high-voltage power system (> 100 kV) [66].

The unit cost of the d.c. supply decreases with increasing voltage and current. A plant is therefore most economical when as many high-current cells as possible are connected in series [67]. Total currents of 450 000 A are achieved. The switches for short-circuiting the cells are designed for 10 000 – 30 000 A and are operated by compressed air, hydraulically, or by spring action. Erosion of the main contacts is dealt with by using replaceable pre-contacts [68]. The contacts are protected from corrosion by installation in vacuum housings.

The current in the bus bars or in anode rods can be measured by means of iron-free transportable equipment with an accuracy of ca. 1 % [69].

5. Mercury Cell Process

The clean separation of chlorine from the cathode products is possible because of the high overvoltage of hydrogen at the mercury electrode. Hydrogen and sodium hydroxide are not produced at the cathode; instead, sodium is produced and dissolves in the mercury as an *amalgam.* The liquid amalgam is removed from the electrolytic cell to a separate reactor, called the decomposer or denuder, where it reacts with water in the presence of a catalyst to form the sodium hydroxide and hydrogen gas. The process may also be used to produce potassium hydroxide by feeding the cell with potassium chloride solution, although this is much less common. The sodium hydroxide is produced from the denuder at a concentration of ca. 50 wt %; the maximum value is 73 wt %. The hydroxide solution is very pure and almost free from chloride contamination.

The process was developed in 1892 almost simultaneously by H. Y. Castner and C. Kellner and used on an industrial scale, although the amount of chlorine produced remained relatively small until 1930, when the rapid growth of the rayon (artificial silk) industry, especially in Germany, increased the demand for pure chloride-free sodium hydroxide solution. At this time, the horizontal high-current cell was developed and output increased rapidly. The development work in Germany was described in the FIAT final reports, published after World War II, and this led to widespread use of the process in Europe and Japan [70]. In the United States, the mercury cell process became more widespread, increasing its share of chlorine production from 3 – 4 % in 1945 to 20 % in 1960, reaching a maximum of 27 % in 1970.

The development of the mercury cell can be followed in the technical data: the cell current increased from 3.4 kA in 1895 to ca. 30 kA in 1945, 200 kA in 1960, and 450 kA in 1970. The current density rose from 2 kA/m^2 in 1950 to the current maximum of 15 kA/m^2. The cell area increased over the same period from ca. 7 m^2 to 37.5 m^2, while the *k*-factor (specific voltage coefficient, see p. 34) was reduced by 50 %.

Since 1972 the importance of the mercury cell has decreased. Increasing concern about the effect of mercury on the environment has led to a considerable increase in the number and variation of statutory regulations that affect the mercury cell process.

In particular, widespread concern about cases of mercury poisoning in Japan, which were not related to the mercury cell process [71], caused the process to be legally banned since 1972. However, conversion to the alternative processes was delayed because of demand for low-chloride sodium hydroxide and because of the anticipated advantages of the rapidly developing membrane cell process. The last remaining mercury cell installations in Japan for NaCl were closed in 1986.

In the other countries, existing mercury cell plants are still in operation, but official regulations and uncertainty about possible further legal restrictions have hindered expansion.

In Europe and the United States great efforts are being made to develop methods of protecting the environment from mercury (see Section 5.3.5). These measures have greatly reduced emissions of mercury into the atmosphere and into wastewater to the extent that the present levels of emitted mercury are negligible in comparison to those arising from natural sources, such as volcanic action, geological erosion, or other nonnatural sources such as fuel combustion or metallurgical processes.

In 1984, the mercury cell process accounted for 45 % of world chlorine production [72]. Since then no new plants using this technology have been built. In the coming decades most of the existing mercury cell plants will be shut down or converted to membrane cell technology. Only plants with speciality products such as extremely pure sodium hydroxide, potassium hydroxide, alcoholates, and dithionites will use the mercury process in future. These plants will meet the highest emission control standards.

5.1. Principles [5]

The cathode reaction

$$Na^+ + e^- + Hg_x \longrightarrow NaHg_x$$

forming sodium amalgam, is followed by the decomposition reaction in a separate reactor

$$2\ NaHg_x + 2\ H_2O \longrightarrow 2\ NaOH + H_2\ (g) + 2\ Hg_x$$

Process Description (Fig. 16). Mercury flows down the inclined base of the electrolytic cell (A). The base of the cell is electrically connected to the negative pole of the d.c. supply. On top of the mercury and flowing cocurrently with it is a concentrated brine with a sodium chloride content of ca. 310 g/L at the inlet. Anodes are placed in the brine so that there is a small gap between the anode and the mercury cathode. The concentration of the amalgam is maintained at 0.2 – 0.4 wt % Na, so that the amalgam flows freely (Fig. 17). The chlorine gas and depleted brine (270 g/L) flow out of the cell, either separately or as a two-phase mixture separated later in the process. The amalgam flows out of the cell through a weir and into the decomposer. The amalgam may be passed through a water wash between the cell and the decomposer to remove traces of sodium chloride. The amalgam flows through the decomposer countercurrent to a flow of softened or demineralized water in the presence of a catalyst to produce sodium hydroxide solution and hydrogen. Stripped of its sodium, the mercury flows out of the lower end of the decomposer and is recirculated through a pump back into the cell.

Anode Reactions. The oxidation of chloride ions to chlorine gas has a standard potential of 1.358 V. In a 300 g/L sodium chloride solution at 70 °C, the reversible reaction potential is reduced to 1.248 V [15, p. 339]. Some side reactions occur, such as

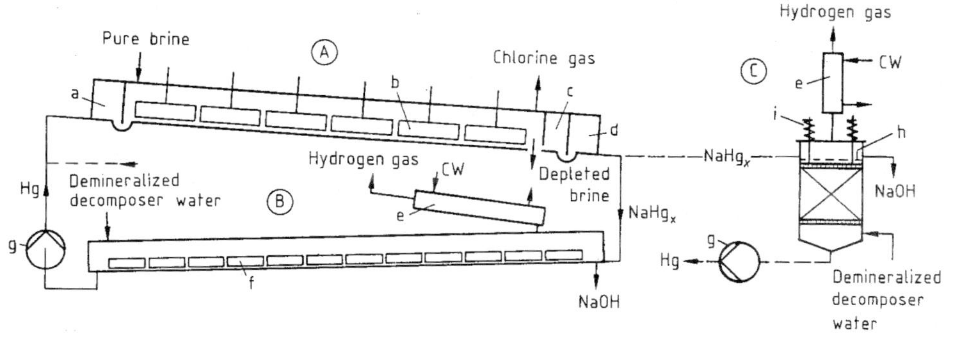

Figure 16. Schematic view of a mercury cell with decomposers
A) Mercury cell: a) Mercury inlet box; b) Anodes; c) End box; d) Wash box
B) Horizontal decomposer: e) Hydrogen gas cooler; f) Graphite blades; g) Mercury pump
C) Vertical decomposer: e) Hydrogen gas cooler; g) Mercury pump; h) Mercury distributor; i) Packing pressing springs

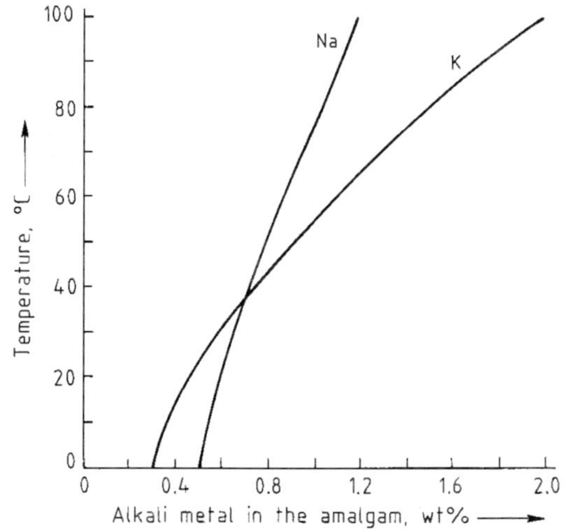

Figure 17. Freezing point curves of sodium amalgam and potassium amalgam

the oxidation of OH^- and SO_4^{2-} ions and the electrochemical formation of chlorate ions. Nonelectrochemical reactions also take place in the region of the anode, such as hypochlorite formation (because of hydrolysis of chlorine) and chlorate formation. All of these side reactions represent a loss of efficiency.

Cathode Reactions. The standard potential of the hydrogen-liberating reaction is 0 V, which is considerably higher than the potential for the formation of 0.2 wt % sodium amalgam, – 1.868 V. However, hydrogen is not liberated at the mercury surface because the reaction is kinetically inhibited. Mainly sodium ions are discharged. At the sodium chloride concentrations used, the reversible potential is reduced by ca. 0.2 V.

Table 4. Electrochemical equivalents f, $kgkA^{-1}h^{-1}$

Element	Element produced	Salt required	Alkali produced
Na	0.8580	2.1810 (NaCl)	1.4923 (NaOH)
K	1.4586	2.7816 (KCl)	2.0931 (KOH)
Cl_2	1.3228 (0.4115 m^3 STP)		
H_2	0.0376 (0.4185 m^3 STP)		

(Exact values of the discharge potential are given as a function of the sodium concentration in the amalgam, the sodium chloride concentration in the brine, and the temperature [73].) Electrochemical side reactions occur: the reduction of chlorine molecules or hypochlorous acid and the liberation of hydrogen gas. In addition, sodium in the amalgam can react directly with free chlorine, or chlorite and chlorate ions can be reduced to chloride by the action of nascent hydrogen at the cathode. All of these side reactions represent a loss of efficiency, normally ca. 2 – 4 % under good operating conditions.

Contamination of the system by heavy metals can lead to a reduction of the hydrogen discharge potential at the mercury cathode, thus increasing hydrogen liberation, and reducing amalgam formation [74]. The hydrogen concentration in the chlorine can increase to the point at which the cell and downstream chlorine handling equipment contains explosive mixtures. The probability of such problems is estimated by a hazard analysis of an existing plant [75] – [77].

The cell system is sensitive to trace quantities of catalysts in the brine, for example, vanadium, molybdenum, and chromium at the 0.01 – 0.1 ppm level or iron, cobalt, nickel, and tungsten at the parts per million level. Magnesium, calcium, aluminum, and barium are also active at the parts per million level.

In addition, relatively high concentrations of sodium in the amalgam (> 0.5 wt%) can cause increased hydrogen evolution in the cells. Potassium chloride electrolysis is considerably more sensitive to both catalysts and high concentration in the amalgam than the sodium chloride process.

Current Efficiency. The theoretical electrochemical equivalents representing the materials produced or consumed in the electrolysis of sodium chloride or potassium chloride brines are given in Table 4. In practice, the yield is ca. 95 – 97 % of the theoretical value, owing to side reactions at the electrodes and in the electrolyte. With activated titanium anodes, the yield is largely independent of the distance between the electrodes.

The decrease in salt concentration Δc is determined by the current I, the brine flow rate M, and the electrochemical equivalent f.

$$\Delta c = fI/M$$

The usual units are c in g/L, f in kg kA^{-1} h^{-1}, I in kA, and M in m^3/h.

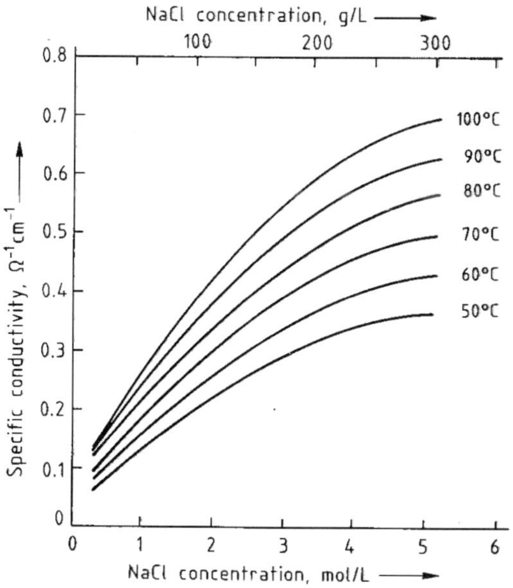

Figure 18. Specific conductivity of sodium chloride solutions

Cell Voltage. The d.c. voltage across the cell circuit is determined by five factors:

1) The reversible decomposition voltage of the salt
2) The overpotentials of the chlorine and alkali metal at the electrodes
3) The voltage drop in the electrolyte
4) Voltage losses in the bus bars, switches, electrical conductors, anode materials, and cathode
5) The operating current density of the cells

Factor 1. The reversible decomposition potential of NaCl under standard conditions is $E° = 3.226$ V (KCl $E° = 3.234$ V). Under the operating conditions $c_{NaCl} = 290$ g/L, $c_{amalgam} = 0.15$ %, and 70 °C, the reversible decomposition voltage is $E = 3.095$ V [78].

Factor 2. The overpotential of chlorine depends on the material and shape of the anodes. At the high current densities (10 kA/m^2) present in modern cell rooms, the overpotential can reach several hundred millivolts, outweighing the effect of concentration changes in the electrolyte (concentration polarization) [79] and the retarding effect that formation of molecular chlorine has on the process of ion discharging [80]. Chlorine gas bubbles cover part of the anode surface, thereby increasing the current density at the free surface. The anode is designed so that the gas bubbles are liberated as quickly as possible. The rapid removal of these gas bubbles from the reaction zone is one of the advantages of titanium anodes over graphite anodes [81].

The overpotential of sodium on the amalgam cathode is caused by the limited diffusion rate of the liberated sodium atoms into the amalgam, but it is small compared to the chlorine overvoltage.

Factor 3. The specific conductivity of sodium chloride solutions increases with concentration and temperature (Fig. 18), but is independent of pH over the range 2 – 11. The brine normally enters the cells at 60 – 70 °C and leaves the cells at 75 – 85 °C. The conductivity of potassium chloride solutions at 70 °C is 30 % greater than that of sodium chloride solutions. Chlorine gas bubbles in the electrolyte

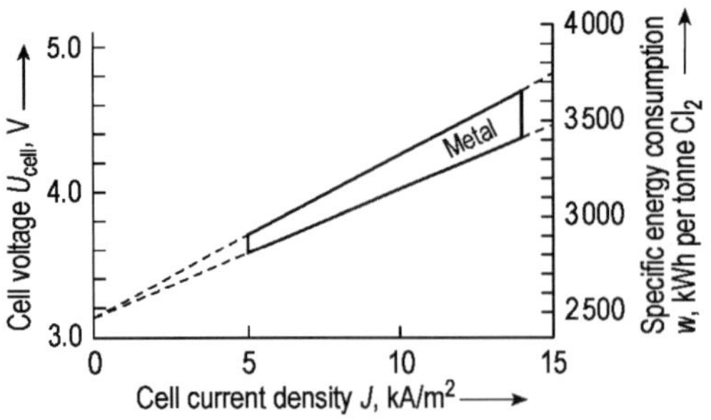

Figure 19. Cell voltage and specific energy consumption per tonne of Cl_2 versus cell current density

increase the resistance between anodes and cathode. Better circulation of the electrolyte in the gap between electrodes allows more rapid removal of gas bubbles, thus reducing the voltage.

Factor 4. The voltage losses in the cell room are minimized by compactly arranging the cells, which shortens the current path. The relatively low conductivity of steel cell bases can be improved by copper or aluminum fittings. These measures also reduce problems caused by magnetic fields, which occur in wide cells [82].

Factor 5. In practice, cell current density and cell voltage have a linear relationship. The slope of the line is termed the specific voltage coefficient or *k*-factor, a useful measure of the specific energy requirement of cells produced by different manufacturers.

The cell voltage is given by $U_{cell} = 3.15 + kJ$, J = current density, kA/m², k = specific voltage coefficient, V m² kA^{-1}.

Computer-controlled cells with activated titanium anodes are run with *k*-factors from 0.085 to 0.11. The corresponding cell voltages at 10 kA/m² are 4.00 – 4.25 V (Fig. 19).

In addition to the d.c. voltages considered above, there are energy losses across the transformer and rectification equipment. All cell installations use a.c. power, which is rectified by silicon diodes in which the energy losses are minimized by operating at greater than 100 V. This voltage is achieved by operating at least 25 cells in series.

Energy Consumption. To operate a cell installation economically, the consumption of d.c. electrical energy per unit mass of product must be minimized. The specific energy consumption *w* is given by

$$w = 1000\ U_{cell}/af$$

where w = kW h/t, a = yield factor or current efficiency, f = electrochemical equivalent, kg kA^{-1} h^{-1}

For example, if the cell voltage U_{cell} is 4.20 V and the current efficiency is 0.970, then ca. 3275 kW h is required to produce 1 t of chlorine + 1.13 t of caustic soda. Since *af* is almost a constant, the specific energy consumption per tonne of chlorine *w* is effectively proportional to the cell voltage. In that case, *w* also depends on the cell current density (see Fig. 19). In the example, w = 3275 kW h corresponds to 10 kA/m².

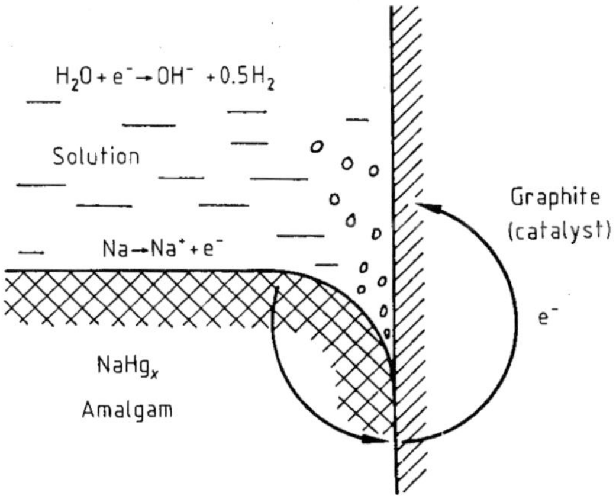

Figure 20. Principle of amalgam decomposition

The total energy requirement per tonne of Cl_2 must also include the transformer and rectifier losses (30 – 40 kW h/t) and the energy requirements of all of the ancillary equipment (120 – 160 kW h/t).

A mathematical model of the cell has been described [83].

Decomposition of the Amalgam. The amalgam is decomposed in horizontal decomposers, alongside or beneath the cell, or more often since ca. 1960 in vertical decomposers or denuders. The energy stored in the amalgam has an emf of ca. 0.8 V. The hydrogen overpotential at the amalgam prevents spontaneous decomposition in contact with water, and a catalyst (depolarizer) must be used. The overall decomposition reaction is

$$2\,NaHg_x + 2\,H_2O \longrightarrow 2\,NaOH + H_2 + 2\,Hg_x$$

and takes place in two stages, first as an anode reaction at the surface of the amalgam

$$2\,Na \longrightarrow 2\,Na^+ + 2\,e^-$$

and then as a cathode reaction on the catalyst surface, where the water is decomposed

$$2\,H_2O + 2\,e^- \longrightarrow 2\,OH^- + H_2$$

Industrial decomposers are essentially short-circuited electrochemical primary cells (Fig. 20). The most common *catalyst* is graphite [*7782-42-5*], usually activated by oxides of iron, nickel, or cobalt or by carbides of molybdenum or tungsten. The hydrogen overpotential on graphite (0.5 – 0.6 V at 2 kA/m^2 and 80 °C) increases with current density and decreases with temperature; therefore, the decomposer should be operated at as high a temperature as possible [84]. Good catalyst material must meet many requirements: resistance to alkali solutions, hydrogen, and mercury; low hydrogen

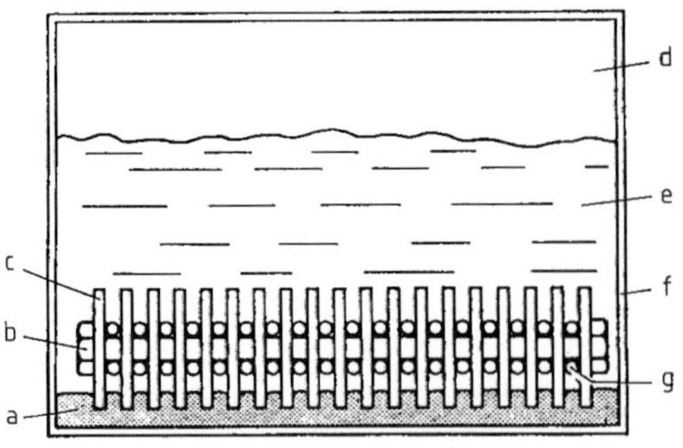

Figure 21. Cross section through a horizontal decomposer
a) Amalgam; b) Bolt; c) Graphite blades; d) Hydrogen; e) Sodium hydroxide solution; f) Decomposer casing; g) Spacers

overpotential; good electrical conductivity; long-lasting activity; wettability by amalgam; and incapability of amalgamation.

Attempts to recover some of the energy stored in the amalgam by creating an electrical circuit by using the catalyst as the anode separated from the amalgam or by using the amalgam electrode with an oxygen gas diffusion electrode have so far had no practical outcome [85].

Horizontal decomposers are ducts with a rectangular cross section, which are installed with a 1 – 2.5 % slope near to or underneath the cells. The amalgam flows in a stream ca. 10 mm in depth, and the sodium content is thereby reduced to < 0.02 wt %. The catalyst consists of graphite blades 4 – 6 mm thick, which are immersed in the amalgam in a lengthwise direction (Fig. 21, also see Fig. 16 B). The water for the reaction, which is softened or demineralized by ion exchange, flows in the direction opposite the amalgam and is removed as 50 % caustic alkali solution. The hydrogen gas is cooled as it leaves the decomposer so that any condensed water and mercury run back into the decomposer. Advantages of the horizontal decomposer are serviceability, simple construction, and a pure product that is low in mercury. However, horizontal decomposers require a greater mercury inventory than vertical decomposers.

Vertical decomposers are designed as towers [86] containing packings of activated graphite spheres or other shapes 8 – 20 mm in diameter. The towers are packed 0.6 – 0.8 m high. The cross section of the tower is 0.35 m^2 per 100 kA of cell current. The amalgam is fed in via an overhead distributor, and the mercury is pumped from the base of the tower back to the cell by a closed centrifugal pump (see Fig. 16 C). The water for the reaction is fed into the base of the tower and flows upward counter to the amalgam. The 50 % caustic alkali solution flows out at the top. The smaller volume of the vertical decomposer leads to higher product temperature because of the greater energy intensity of the system. Cooling the hydrogen is essential. Compared with the horizontal decomposer, the amount of space required is small, and the mercury inventory is small, but the caustic alkali contains more mercury.

In alternative decomposition reactions, other products may be obtained from the amalgam in place of sodium or potassium hydroxide solutions [10, p. 518], [87]: sodium sulfide from sodium polysulfide solution, alcoholates from alcohols, sodium dithionite from sodium hydrogen sulfite, hydrazobenzene or aniline from nitrobenzene, adiponitrile from acrylonitrile, and alkali metals by distillation.

5.2. Mercury Cells

During the first decades after the rocking cells of CASTNER and KELLNER were first commissioned, considerable efforts were made to develop suitable materials for the cells and the anodes. A large number of cell configurations were tested, resulting in the development of the continuous cell. Since 1950, the cell areas and the specific load were increased considerably.

In 1972, the changeover from graphite to metallic anodes began, with a parallel development of computer monitoring and control, leading to improved short-circuit protection and a reduction of the specific energy consumption by computer-controlled anode adjustment, of great significance in view of the drastic increase in electricity costs in the late 1970s. In the years following 1972, producers operating the electrolysis plants also concentrated on the development and installation of devices to reduce mercury emissions.

The cells currently available possess a number of common features. The mercury flows over a steel base that has a slope of 1.0 – 2.5 %. The flanged side walls are lined with rubber. The cell covers are mostly steel, lined with rubber or titanium on the underside, but they may also be rubberized fabric. The anodes, today almost always of activated titanium, hang in groups from carrying devices that can be varied in height manually, hydraulically, or by motor-operated lifting devices. Each cell can be short-circuited externally by a switch. The cell bus bars are usually copper. The anodes are protected from internal short circuits by means of electronic monitoring systems. The size of the cells can be varied within a broad range to give the desired chlorine production rate. Computer programs optimize the cell size, number of cells, and optimum current density as a function of the electricity cost [88] and capital cost.

For comparison, a list of cells manufactured by leading engineering firms and cell characteristics is given in Table 5. Cathode surface areas are ca. 17 – 30 m^2, and nominal currents are ca. 170 – 300 kA [32, p. 204].

5.2.1. Uhde Cell

The Uhde cells (Fig. 22, also see Fig. 25) are available with a cathode surface area 4 – 30 m^2 for chlorine production rates 10 – 1000 t/d for the complete cell installation. The brine flows in via an inlet box fitted with two pipes for the removal of chlorine. The weak brine is removed at the end of the cell. The solid cover is fixed to the side walls by clamps. The anodes are suspended in groups in carrying frames supported near the cells on transverse girders with lifting gear. The anode rods are raised and lowered

Table 5. Characteristics of modern mercury cells

Characteristic	Manufacturer				
	Uhde	De Nora	Olin – Mathiesen	Solvay	Krebs Paris
Cell type	300 – 100	24M2	E 812	MAT 17	15 KFM
Cathode area, m^2	30.74	26.4	28.8	17	15.4
Cathode dimensions, $l \times b$, m^2	14.6×2.1	12.6×2.1	14.8×1.94	12.6×1.8	9.6×1.6
Slope of cell base, %	1.5	2.0	1.5	1.7	
Rated current, kA	350	270	288	170	160
Max. current density, kA/m^2	12.5	13	10	10	10.4
Cell voltage at 10 $kA/m^2 \cdot V$	4.25	3.95	4.24	4.10	4.30
Number of anodes	54	48	96	96	24
Stems per anode	4	4	2	1	4
Number of intercell bus bars	36	32	24	24	12
Quantity of mercury per cell, kg	5000	4550	3800		1650
Energy requirement per tonne of Cl_2, kW h d.c.	3300	3080	3300	3200	3400

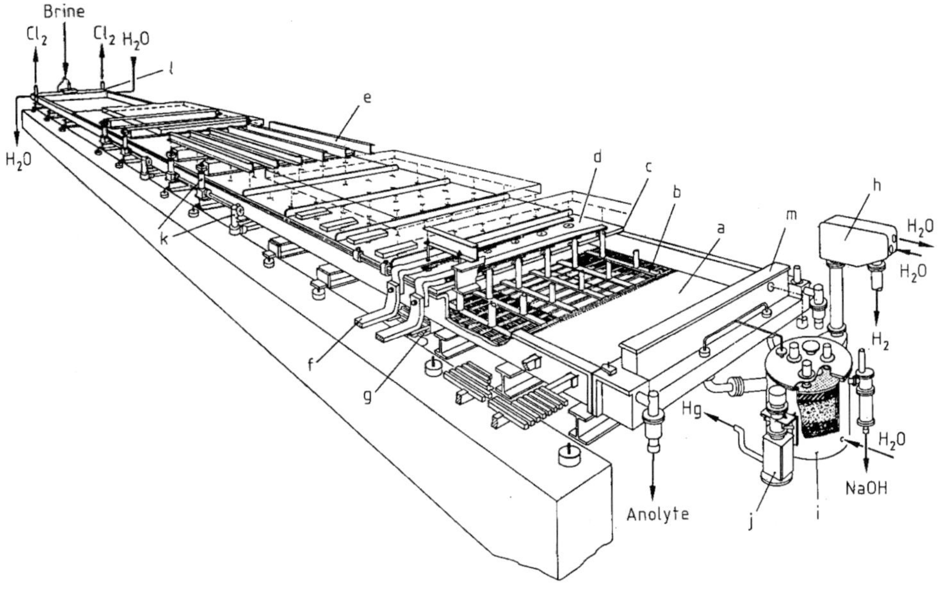

Figure 22. Uhde mercury cell
a) Cell base; b) Anode; c) Cover seal; d) Cell cover; e) Group adjusting gear; f) Intercell bus bar; g) Short-circuit switch; h) Hydrogen cooler; i) Vertical decomposer; j) Mercury pump; k) Anode adjusting gear; l) Inlet box; m) End box

within a bellows seal. Short copper bus bars between the cells also serve for shunt measurement of the anode currents. The electric current is brought in above the cell covers via flexible copper straps that run immediately above the anode rods and are bolted to them. The compressed-air switches are situated under the cells. The cell bottom is usually a current conductor when cells are short-circuited, but in wide, heavily loaded cells the cathode current is carried by copper bus bars to prevent the

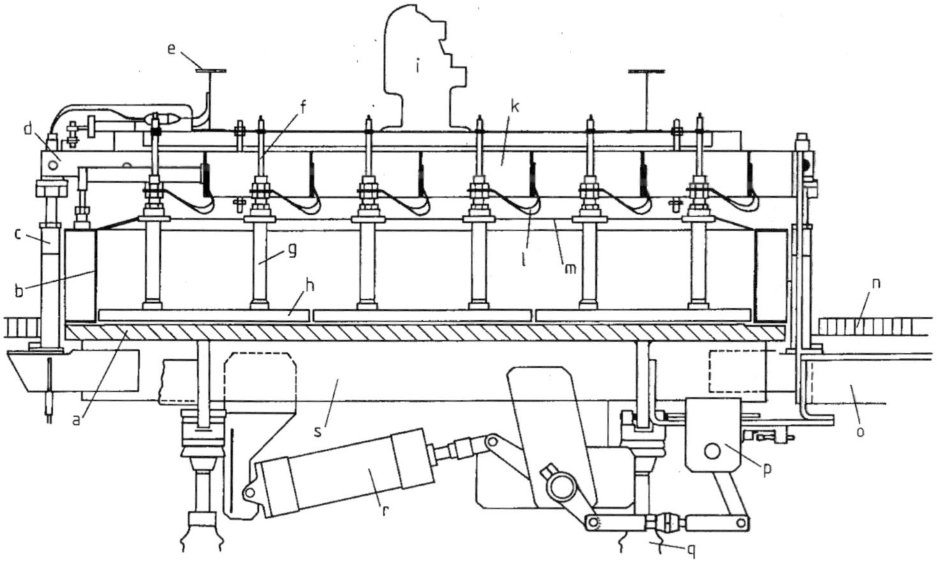

Figure 23. Cross section through the De Nora mercury cell
a) Cell base (steel); b) Side wall (rubber-lined steel); c) Lifting gear; d) Transverse support; e) Lengthwise support; f) Anode carrier; g) Anode rod; h) Anode surface; i) Adjusting motor; k) Bus bar; l) Flexible anode current strap; m) Multilayer cell cover; n) Service walkway; o) Intercell bus bar; p) Switch; q) Insulator; r) Switch drive; s) Support

occurrence of strong magnetic fields, which could interfere with the amalgam flow. The automatic equipment for protection and adjustment of the anodes depends on the shunt measurement of the currents and is controlled by a central computer. In this way, an optimum *k*-factor is selected for each cell. The vertical decomposers are provided with hydrogen coolers and are situated at the end of each cell. The amalgam flows into the decomposer under the force of gravity [89].

5.2.2. De Nora Cell

The size of the De Nora cell (Fig. 23, also see Fig. 26) varies from 4.5 to 36 m^2, corresponding to electric currents from 45 to 400 kA. The cover is a flexible multilayer sheet of elastomer spread over the cell trough. This cover is supported by the anode rods and seals them. The DSA anodes (see Section 8.1) are held rigidly in strong carrying frames, which are automatically adjusted by electric motors. Individual anode adjustment is not provided. The anode rods are individually connected by flexible copper straps to the anode bus bars. The cathode current is carried by copper bus bars. Devices for the improvement of brine circulation and gas removal within the cells reduce specific energy consumption. Consequently, the reduction in brine concentration can be increased from the usual 35 – 40 g/L to 60 – 70 g/L, and the brine circulating rate can be reduced by ca. 40 %. Separate outlets are present at the inlet box for the normal chlorine gas production and the weak chlorine gas produced during start-up. The graphite catalyst in the vertical decomposer is activated with molybdenum.

5.2.3. Olin – Mathieson Cell

The special feature of the Olin – Mathieson cell lies in the system of mounting and adjusting the anodes. Above each row of anode rods, a U-shaped copper or aluminum bus bar also serves to support the anode lifting gear. The anode rods are bolted to the U-shaped bus bar. The anodes are adjusted as a group, either manually or by a remote computer with the remote computerized anode adjuster (RCAA) system. The currents are measured independently of the cell potentials by means of reed contacts [90].

5.2.4. Solvay Cell

The bus bars in the Solvay cells are made primarily of aluminum. Above the cells is a cover that also serves as a convenient walkway, giving access to the anode rods. The titanium anodes are specially coated and are automatically adjusted by computer. The tall vertical decomposers are located under the cells.

5.3. Operation

The aspects of the operation of mercury cells that typically differ from those of the other processes are the brine circulation system, the cell room, treatment of the products, measurement and control, and reduction of mercury emissions.

5.3.1. Brine System

A typical brine circulation system for the mercury cell process is shown in Figure 24. In the cells the sodium chloride concentration of the brine is reduced by 35 – 60 g/L to 260 – 280 g/L at 70 – 85 °C. To avoid mercury emissions into the air, the resaturators are generally closed vessels. The mercury cathode is very sensitive to poisoning by heavy metals; therefore, a test [91] has been developed that allows rapid determination of the suitability of any particular salt or brine.

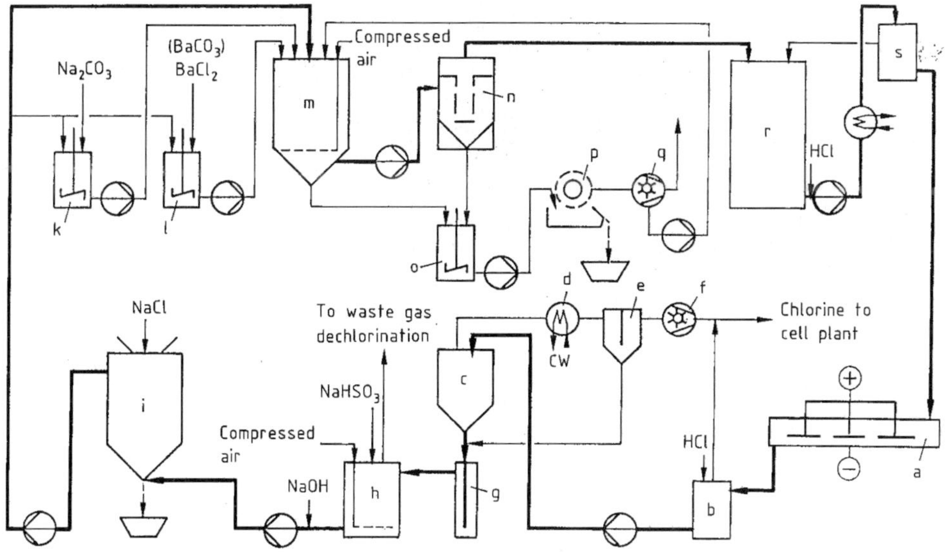

Figure 24. Schematic diagram of a brine circulation system in the mercury cell process
a) Electrolysis cell; b) Anolyte tank; c) Vacuum column dechlorinator; d) Cooler; e) Demister; f) Vacuum pump; g) Seal tank; h) Final dechlorination; i) Saturator; k) Sodium carbonate tank; l) Barium chloride tank; m) Brine reactor; n) Brine filter; o) Slurry agitation tank; p) Rotary vacuum filter; q) Vacuum pump; r) Brine storage tank; s) Brine supply tank

5.3.2. Cell Room

The cells are usually situated in a building (Fig. 25), although sometimes they are erected in open air (Fig. 26). Figure 27 shows a bird's eye view, and Figure 28 shows a cross section of cell room. The cells are arranged parallel to each other so that bus bars and supply lines are kept short. The cells stand on supporting structures and are insulated to prevent shorting to the earth. The transformer and rectifiers are situated at one end of the room, and the cell service and repair area is at the opposite end. Ancillary equipment is installed near the cell room in a spillage containment area.

Cell floors, gangways, and spillage containment areas are constructed with smooth, sloping floors so that any mercury can be easily recovered or wash water can be conveniently collected for treatment. The supply pipes run under the cells and are connected to them by flexible, insulating connections. The heat given off by the cells and the decomposers is removed by a ventilation system.

The plant is operated with continuous 24 h/d supervision and control. An additional day-shift team carries out anode changes, repairs, and cleaning.

Occupational Health. Anyone working in the cell area must undergo regular health checks. Euro Chlor has prepared a Code of Practice "Control of Worker Exposure to Mercury in the Chlor-Alkali Industry" [92]. The U.S. Chlorine Institute has released guidelines "Medical Surveillance and Hygiene Monitoring Practices of Worker Exposure

Figure 25. Cell room: Uhde mercury cells

Figure 26. Open-air cell room: De Nora cells

to Mercury in the Chlor-Alkali Industry" [93]. The U.S. Environmental Protection Agency has established 18 rules relating to cleanliness of the cell room [94]. Adherence to these rules eliminates any danger to the health of personnel caused by mercury. The maximum allowable concentration or threshold limit value (TLV) of mercuryindentry relevance="low">threshold limit value (TLV) of mercury in the atmosphere in Western Europe and in the United States is between 0.025 and 0.100 mg/m^3.

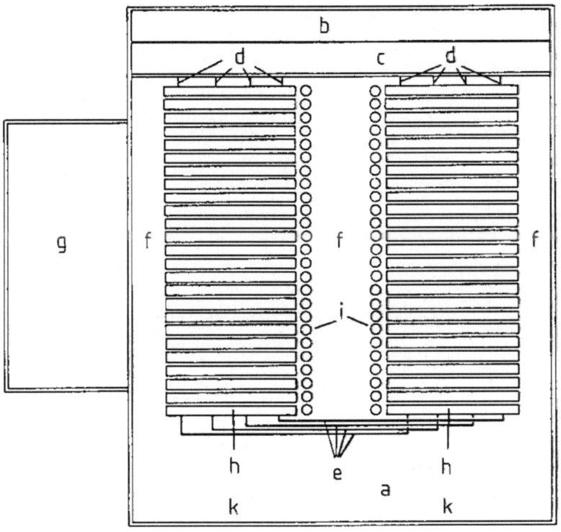

Figure 27. Mercury cell room (bird's-eye view, schematic)
a) Cell room; b) Transformer room; c) Rectifier room; d) Bus bars; e) Turnaround bus bars; f) Service walkways; g) Ancillary equipment; h) Electrolysis cells; i) Vertical decomposers; k) Cell assembly and maintenance area

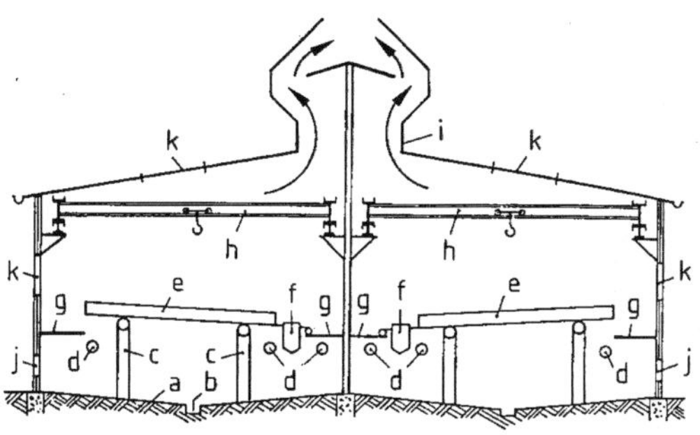

Figure 28. Mercury cell room (cross section, schematic)
a) Basement floor; b) Floor drains; c) Cell supports with insulators; d) Supply pipes; e) Cells; f) Decomposers; g) Service walkways; h) Crane; i) Ridge ventilator; j) Ventilation air supply; k) Windows/lighting

5.3.3. Treatment of the Products

Chlorine. See Chapter 11.

Hydrogen. The treatment of the hydrogen gas leaving the decomposer is described in Chapter 4. It must pass special equipment for the removal of the traces of mercury before it is used (see p. 46).

Sodium Hydroxide Solution. The great advantage of the mercury cell process is that very pure sodium hydroxide solution is produced (see Table 20) at a suitable concentration. The chloride content is only 5 – 50 mg/kg.

Sodium hydroxide from the decomposer usually has a concentration of 50 % and a temperature of 80 – 120 °C. It passes through rubber-lined steel pipe work to nickel or Incoloy coolers, where it is cooled to 40 – 60 °C. Any particles of graphite from the decomposer or traces of mercury are effectively removed by centrifuges, candle filters, or precoated leaf filters (Fig. 29).

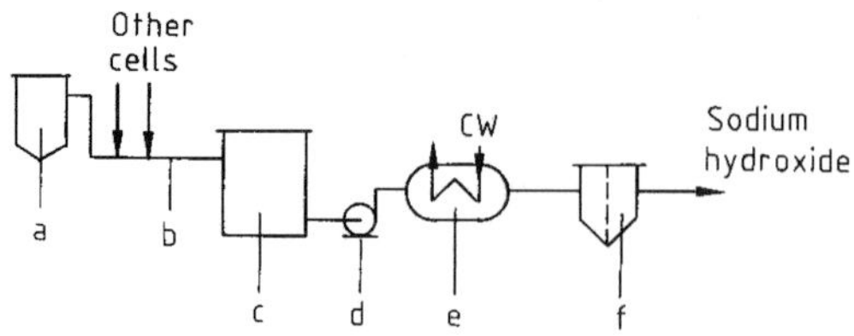

Figure 29. Processing of sodium hydroxide solution from the amalgam decomposer
a) Vertical decomposer; b) Collection main; c) Collecting tank; d) Pump; e) Cooler; f) Mercury removal filter

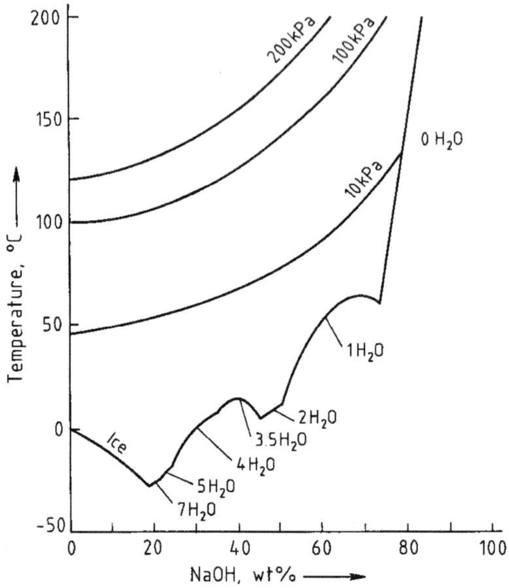

Figure 30. Freezing and boiling point curves of sodium hydroxide solutions

The freezing-point and boiling-point curves of sodium hydroxide solutions are shown in Figure 30. The phases separating from the solution, i.e., ice, hydrates, and NaOH, are indicated along the freezing-point curve. Sodium hydroxide is supplied to consumers as aqueous solution, solid block, flakes, prills, or powder. For the processes involved and for uses (see Chapt. 9.3).

5.3.4. Measurement

The condition of the brine, the cells, and the products must be continuously and carefully monitored, since even small deviations from the correct conditions can increase the hydrogen concentration in the chlorine. The measuring operations are mostly automatic: critical limits are chosen and if these limits are exceeded, alarms are set off.

Cell Operating Conditions. The sodium concentration in the amalgam is determined at the cell inlet (max. 0.05 wt %) and outlet (max. 0.45 wt %).

A 20-g sample of the amalgam is reacted with 30 wt % aqueous sulfuric acid in an absorption pipette. The evolution of 1 cm^3 of gas corresponds to 0.01 wt % Na.

A portable analytical and recording apparatus is available that works electrochemically [95].

If the mercury pump stops, the steel cathode base of the cell is exposed to electrolyte, and hydrogen evolves to form an explosive mixture with the chlorine in the cell.

Failure of the mercury pump or mercury flow automatically short-circuits the cell. Mercury flow failure is detected by monitoring the mercury level at the lowest part of the mercury circulation system or in the inlet box, by direct flow measurement [96], or by loss of pressure at the pump delivery.

The motors for the mercury pumps, the chlorine absorption plant, and the most vital control equipment are all provided with an emergency power supply, ensuring safe shutdown of the plant if a power failure occurs. The installation is protected by a complex system of interlocks so that failure of important equipment, such as the chlorine compressor, shuts down the rectifiers.

A large number of systems have been developed for the protection of cells from short circuits. Titanium anodes are destroyed by short circuits and must be raised before any contact with the amalgam takes place.

The operation of the monitoring system depends on magnetic-field current measurement for individual anodes [90], [97], [98] or on shunt measurement of the supply bus bars [99]. Monitoring is achieved by comparison of the anode – cathode voltage of different cell sections [100] or by following the conductivity of the brine in the electrode gap [101].The signals from the instruments are fed into central computers or local microprocessors at each cell [102] that control the anode lifting gear. The mercury inventory in each cell may be measured by a radioactive tracer technique once a year without affecting cell operation [103].

Products. The concentration of the sodium hydroxide solution is determined from its density, and the purity its checked by titration to determine hydroxide, carbonate, and chloride contents. The purity of the water for the decomposer is determined from its conductivity.

The oxygen content of the hydrogen gas is determined from the magnetic susceptibility, oxygen being paramagnetic.

5.3.5. Mercury Emissions [71]

Any chlor-alkali plant up to modern technical standards is not a hazard to the environment. The residual emissions to water and air are ecologically acceptable according to current knowledge. The mercury in the electrolytic cells circulates in a closed system. All materials that come into contact with the mercury — equipment, products, auxiliary chemicals, wash water, waste gases, other waste materials — may become contaminated with mercury, and must be treated before release to the environment or must be safely deposited. For the exact measurement of these trace amounts and for control of the effectiveness of the measures to reduce the emissions, analytical methods have been developed with sensitivities in the microgram region [104].

Many countries have set legal limits on emissions in waste air and water. The limits on the products of a chlor-alkali plant may depend on their end use, e.g., drinking water treatment or food processing. The sources of contamination are listed, and means of reducing them are described: Mercury cells are sealed vessels, and the products are conveyed in closed pipes. The cell rooms must have smooth joint-free floors with easily cleaned drainage surfaces and irrigated collection gutters (see Fig. 28). Spilled mercury is immediately washed away with water into collecting tanks or sucked up with a vacuum system.

Control of mercury loss is only possible if the mercury content of all the cells is known exactly. The gravimetric and volumetric methods formerly used were cumbersome and led to additional mercury emissions, disadvantages that are avoided by a radioactive tracer method.

Mercury in Products. Hot, moist *chlorine* leaving the cell contains small amounts of mercuric chloride. This is almost completely washed out in the subsequent cooling process and may be fed back into the brine with the condensate. In the cooled and dried chlorine gas, there are only minute traces of mercury: 0.001 – 0.01 mg/kg.

The equilibrium mercury concentration in *hydrogen gas* is a function of temperature and pressure. The mercury concentration (mg of Hg per m^3 of H_2 at 101.325 MPa) increases rapidly with temperature:

t, °C	0	20	40	60	80	100
c, mg/m^3	2.36	14.1	66.1	255	836	2404

Subjecting the hydrogen gas to pressure lowers the mercury content of the resulting product gas at atmospheric pressure. For example, at 5 °C

$$c_{Hg}\,p = 0.37\,\mathrm{Pa\,kg\,m^{-3}}$$

c_{Hg}, concentration of mercury in hydrogen gas at atmospheric pressure, mg of Hg per m^3 of H_2

p, pressure to which the hydrogen gas is subjected, MPa

When the mixture is cooled to 2 – 3 °C, the mercury concentration is reduced to ca. 3 mg/m^3 at standard pressure. This mercury content can be reduced by compressing and further cooling, adding chlorine to form mercurous chloride (calomel), which is collected on rock salt or similar material in a packed column, washing with solutions containing active chlorine, or by adsorption on activated carbon impregnated with sulfur or sulfuric acid, leaving a mercury concentration in hydrogen of

0.002 – 0.03 mg/m^3. The highest purity can be achieved by adsorption on copper/aluminum oxide or silver/zinc oxide, < 0.001 mg/m^3.

Centrifugation or filtration in candle filters or in disk filters precoated with charcoal gives *sodium hydroxide solutions* containing mercury concentrations of < 0.05 ppm (mg/kg of 50 % caustic soda).

The circulating brine contains mercury concentrations of 2 – 20 mg/L. Mercury emissions from the brine system can occur through losses of brine into the wastewater, by brine vaporization in the resaturators, or by disposal of the residues from the brine purification filter. These emissions are minimal at a chlorine concentration < 30 mg/L, giving a redox potential > 500 mV vs. NHE. Under these conditions mercury remains dissolved in the brine as a mercury chloride complex even if the brine is alkaline.

Mercury in Wastewater. Mercury-containing wastewater has several sources:

1) The process, e.g., condensate and wash liquor from treatment of chlorine, hydrogen, and brine; stuffing-box rinse water from pumps and blowers; brine leakages; ion-exchange eluate from process-water treatment
2) Cell cleaning operations
3) Cleaning of floors, tanks, pipes, and dismantled apparatus

The amount of wastewater can be reduced by separately disposing of the cooling water and process water and by feeding the condensate back into the brine, provided the water balance allows this. A wastewater rate of 0.3 – 1.0 m^3 per tonne of chlorine is achievable.

There are various methods of making wastewater suitable for discharge:

1) Chemical removal of mercury by reducing any compounds to the metal with hydrazine or sodium borohydride or by precipitating mercuric sulfide with thiourea or sodium sulfide. The mercury metal or sulfide is then filtered off.
2) Oxidation of the mercury by chlorine, hypochlorite, or hydrogen peroxide and adsorption on an ion-exchange medium. Elutriation is done with hydrochloric acid, which is then used to acidify the brine [105].

The Clean Water Act of 1972 (United States) demands the use of the "best available technology economically achievable." Since 1982 each plant has been limited to a maximum of 0.1 g of Hg per tonne of chlorine averaged over 30 d measured at the outlet of the wastewater treatment plant.

In Western Europe, an EC directive has been issued on the subject of the mercury content of wastewater from chlor-alkali plants, following various earlier agreements such as the Rhine protection agreement, the EC guidelines concerning the protection of natural waters, and the Paris Convention [106]. This directive requires plants with circulating brine systems to have a limit of 1.0 g of Hg per tonne of chlorine produced.

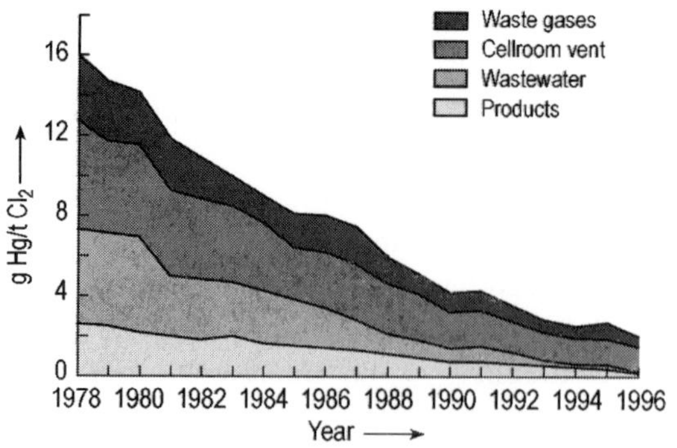

Figure 31. Mercury emissions from the European chlor-alkali industry

Mercury in Process Air. Air from the process, for example, the cell end box ventilation system, vents from liquid collection tanks (caustic, wastewater), from the vacuum cleaning system, or from the distillation unit for mercury-contaminated wastes can be treated to remove mercury by the methods used for hydrogen.

Ventilation of the Cell Room. The heat produced during electrolysis requires that the air must be changed 10–25 times per hour, depending on the type of building. Mercury spillage can occur during essential operations involving cells or decomposers, for example, opening the cells for anode changing or cleaning, assembling or dismantling equipment, or replacing defective pipes. Spillage leads to small losses in the exhaust air owing to the vapor pressure of mercury. In addition, products that contain mercury, such as the sodium hydroxide solution, hydrogen, or process waste air, can escape via faulty seals in pipes and equipment, leading to emissions. Closed cell construction and special care in handling mercury, i.e., good housekeeping, by adhering to the EPA rules or the Code of Practice "Mercury Housekeeping" [107], keep the mercury concentrations and, hence, emissions below the allowable work place concentrations (MAK and TLV) [108].

Purification of large volumes of waste air containing mercury in very low concentrations is not effective. In the United States, the upper limit for the emission of mercury in process waste air and hydrogen is 1 kg per day per facility, and for ventilation air it is 1.3 kg per day per facility [18, p. 372]. In Western Europe the Parcom Decision 90/3 [109] requires a standard of 2 g Hg/t of chlorine capacity for emissions to the atmosphere from existing plants.

Mercury in Residues. Mercury-containing residues include brine filter slurry, spent decomposer catalyst, discarded cell components, residues from the purification of products, waste material from rinsing media, adsorption materials, ion-exchange media, etc. Mercury can be recovered from these materials by distillation in closed retorts. The residues after distillation must be disposed at special sites. Mercury in safely deposited wastes is not considered an emission to the environment.

Summary. The continued efforts of all producers have led to a steady decrease in mercury emissions over the years; for example, in Western Europe from 16 g in 1978 to 2 g Hg per tonne of chlorine capacity in 1996, as shown in Figure 31 [110]. With this low emission level, the contribution of the chlor-alkali industry to the total natural and anthropogenic mercury emissions is less than 0.1 % [112]. Euro Chlor is developing a BAT (best available techniques) for reducing mercury emissions from existing mercury-based plants, the application of which will ensure that in 2007 no plant emits more than 1.5 g Hg per tonne of chlorine capacity to air, water, and products [110].

6. Diaphragm Process

The commercial production of chlorine by electrolytic processes began in Europe and the United States in the 1890s. Early cells of the bell-jar type had no diaphragm and relied on the flow of anolyte toward the cathode to prevent the hydroxide ion from back-migrating toward the anode. This method had limited capacity because gas evolution caused mixing and loss of efficiency. The Griesheim cell, another early design, used porous cement as the diaphragm.

E. A. LE SUEUR is credited with the design of a cell incorporating a percolating asbestos diaphragm, which is the basis for all diaphragm chlor-alkali cells currently in use. When brine is caused to flow into the anolyte and subsequently through the diaphragm into the catholyte, continuous operation with much improved efficiency is obtained. This Le Sueur cell, and the similar Billiter cell developed in Germany, incorporated a horizontal asbestos sheet as the diaphragm. During the 1920s, the Billiter cell became the most widely used cell in the world; a few are still in operation today.

Following the invention of synthetic graphite, numerous cells were developed. These fall into three basic types:

1) Rectangular vertical electrode cells
2) Cylindrical vertical electrode cells
3) The vertical electrode bipolar filter press cell developed by Dow Chemical

In 1913, C. W. MARSH developed a cell with finger cathodes and side-entering anodes and cathodes, which greatly increase the electrode area per unit of floor space. About 1928, KENNETH STEWART of Hooker Chemical (now Occidental Chemical Co.) developed a method of depositing asbestos fibers onto the cathode by immersing the cathode in a slurry of asbestos fibers and applying a vacuum. All significant installations of diaphragm cells currently in operation are derived from that development [2].

All diaphragm cells produce cell liquor that contains ca. 11 wt% caustic soda and 18 wt% sodium chloride. To market the caustic soda, its concentration must be increased to 50%. During the evaporation and cooling processes, the salt becomes less soluble in the stronger caustic, and at 50% NaOH the NaCl concentration is ca. 1%.

6.1. Principles

The principles needed to understand the efficient operation of the diaphragm process involve the current efficiency, cell voltage, power consumption, and optimization of the operating conditions [113], [114].

The reaction at the positively charged anode is the same for all three chlor-alkali processes

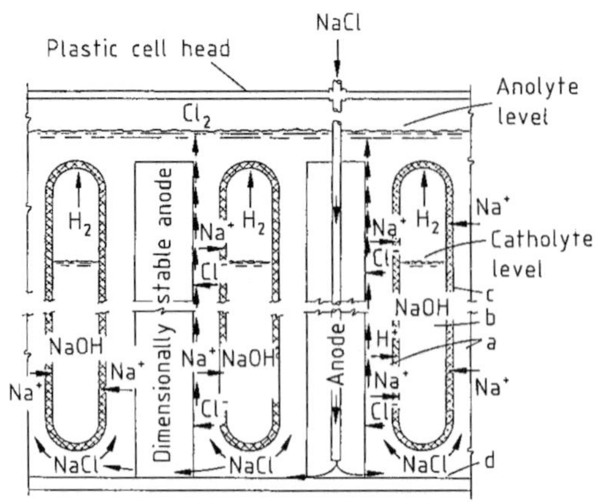

Figure 32. Basic chemical reactions within the cell
a) Anode compartment; b) Cathode compartment; c) Deposited diaphragm on cathode tubes, rims, and end screens

$$2\,Cl^- \longrightarrow Cl_2 + 2\,e^-$$

The reaction at the negatively charged cathode of the diaphragm cell is

$$2\,H_2O + 2\,e^- \longrightarrow H_2 + 2\,OH^-$$

Figure 32 is a cutaway view of a diaphragm cell that shows the orientation of the various parts of the cell and the various reactions that take place. Figure 32 also shows the location of the diaphragm, which is deposited on the outside of the cathode screen and which separates the cell into two compartments, one containing the anodes and one containing the cathode. The sodium chloride solution (brine) enters the anode compartment and completely covers the anodes and the cathode tubes or fingers. The chlorine leaves the cell through an outlet in the cell head. The anolyte flows through the diaphragm into the cathode compartment because of the difference in liquid level between the two compartments. The catholyte is a solution of sodium chloride and sodium hydroxide because a portion of the water is converted to hydroxide at the cathode. The hydrogen produced at the same time leaves the cell through an outlet on the cathode. The solution of sodium chloride and sodium hydroxide overflows the cell through a level control pipe on the cathode, and is then commonly called cell liquor.

Current Efficiency. Current efficiency is defined as the amount of product actually produced divided by the amount of product that theoretically should have been produced on the basis of the amount of direct-current electrical energy input. The current efficiency is never 100 % because of side reactions. The efficiency of a diaphragm cell is usually based on the chlorine production.

The side reactions that lower the efficiency are a result of chlorine that enters the catholyte compartment or of hydroxide ions that enter the anolyte compartment. The amount of chlorine that enters the catholyte compartment is small. The majority of the efficiency losses in a diaphragm cell are due to migration of hydroxide ions from the catholyte through the diaphragm and into the anolyte. This back migration takes place because the negatively charged hydroxide ions are attracted to the positively charged anodes and because of the hydroxide ion concentration gradient across the diaphragm. This migration of hydroxide ions through the diaphragm is in equilibrium with the opposing flow of brine through the diaphragm.

Three factors control the migration of the hydroxide ions into the anolyte:

1) Concentrations of hydroxide and chloride ions at the cathode side of the diaphragm
2) Flow rate of brine through the diaphragm
3) Condition of the diaphragm

These three factors in dynamic equilibrium determine the efficiency of the cell.

Factor 1. The higher the concentration of hydroxide ions in the catholyte, the larger the concentration gradient across the diaphragm, and the higher the probability of hydroxide ions crossing through the diaphragm. As a result, changing cell liquor strength strongly affects cell efficiency. The concentration of chloride ions in the catholyte also affects cell efficiency because some of the chloride ions migrate in place of hydroxide ions.

Factor 2. Decreasing brine flow rate to a cell increases the conversion of sodium chloride to sodium hydroxide and raises the hydroxide concentration in the catholyte because of reduced overflow from the cell. The decreased flow rate of brine through the diaphragm allows increased migration of hydroxide ions into the anolyte. These factors decrease cell efficiency.

Factor 3. The condition of the diaphragm is extremely important. Nonuniformity in the diaphragm results in high flow rates of brine through thin or loosely compacted areas and low flow rates through thick or compacted areas. In the areas where there is a low brine flow rate, back migration of hydroxide is increased.

The degree of inefficiency in a cell is indicated by the two products of the side reactions, oxygen in the chlorine and sodium chlorate in the cell liquor. Oxygen in the chlorine gas is the result of hydroxide ions that migrate through the diaphragm into the anolyte, where they are oxidized:

$$2\ OH^- \longrightarrow 1/2\ O_2 + H_2O + 2\ e^-$$

Sodium chlorate in the cell liquor is a result of hydroxide ions that migrate through the diaphragm into the anolyte and react with chlorine before reaching the anode:

$$3\ Cl_2 + 6\ NaOH \longrightarrow NaClO_3 + 5\ NaCl + 3\ H_2O$$

Equations. The simplest equations for calculating cell efficiency are based on the masses of products produced per unit of electrical input. Theoretically, 1.492 kg of sodium hydroxide and 1.323 kg of chlorine are produced per kiloampere-hour. It then follows that

$$\text{Cathode efficiency, \%} = (\text{kg of NaOH} \times 100)/(Q \times 1.492 \times \text{the number of cells})$$

JOHN RYLANDS
UNIVERSITY
LIBRARY OF
MANCHESTER

Anode efficiency, % = (kg of $Cl_2 \times 100)/(Q \times 1.324 \times$ the number of cells)

where Q is the quantity of electricity in kA h

Unfortunately, the production of a single cell cannot be measured with sufficient accuracy to give meaningful results. To get around this problem, the chlorine industry uses an equation based on the analysis of the chlorine gas, the cell liquor, and the anolyte:

$$\% CE = [\% Cl_2 \times 100]/[\% Cl_2 + 2(\% O_2) + (\% Cl_2 \times anox \times F)/c_{NaOH}]$$

where

% CE	= anode current efficiency, %
% Cl_2	= percent chlorine in cell gas (air free)
% O_2	= percent oxygen in cell gas (air free)
anox	= oxidizing power of anolyte expressed as grams of $NaClO_3$ per liter
c_{NaOH}	= NaOH concentration in the cell liquor, g/L
F	= conversion factor

The denominator is the amount of chlorine produced plus the amount of chlorine consumed in the side reactions. This is equivalent to the amount of chlorine that could have been produced theoretically from the input of current. The conversion factor F is the product of a volume factor, an electric field factor, and a stoichiometric factor. In practice, it is a function of cell liquor strength.

The *SIX equation* is a practical alternative to the previous equation and is often used with computers linked to a gas chromatograph and an automated cell liquor analyzer. The SIX equation is

$$\% CE = [\% Cl_2 \times 100]/[\% Cl_2 + 2(\% O_2) + (\% Cl_2 \times 6 \times c_{NaClO_3})/c_{NaOH}]$$

This equation also accounts for chlorine lost to the anolyte. However, it approximates the oxidizing potential of the anolyte with the concentration of chlorate in the cell liquor and assumes a fixed conversion factor from anolyte concentration to catholyte, namely *SIX*. The SIX equation approximates the standard equation within 0.5 %.

Cell Voltage. The voltage of a cell is the sum of five component voltages: anode potential, cathode potential, cell structure voltage drop, diaphragm voltage drop, and anolyte – catholyte voltage drop. The anode and cathode potentials are sums of the reversible voltages, which are the thermodynamic minimum amounts of work to cause the reactions to take place, and the overpotentials, which are the additional voltages required for nonreversible kinetics. The cell structure voltage drop includes the voltage losses in the cathode, anodes, intercell bus, and all other connectors in the cell. The sum of the diaphragm voltage drop and the anolyte – catholyte (brine) voltage drop is the potential between the electrodes. All of these voltages are functions of current density. Table 6 shows how cell voltage is strongly affected by cell current. Cell

Table 6. Typical voltage distribution [a]

Component voltages, V	Current density, kA/m^2				
	1.24	1.55	1.86	2.17	2.48
Anode potential [b]	1.30	1.30	1.30	1.30	1.31
Cathode potential [b]	1.12	1.13	1.15	1.16	1.17
Structure loss [c]	0.11	0.14	0.17	0.20	0.22
Brine loss	0.11	0.15	0.19	0.23	0.27
Diaphragm loss	0.24	0.31	0.36	0.41	0.47
Intercell bus	0.02	0.02	0.03	0.03	0.03
Total	*2.90*	*3.05*	*3.20*	*3.33*	*3.47*

[a] OxyTech MDC-55 cell with Modified Diaphragm and expandable anodes. Conditions: anolyte temperature 93 °C, anolyte NaCl concentration 250 g/L, catholyte NaOH concentration 130 g/L.
[b] Potential vs. NHE.
[c] Includes anode base, anodes, cathode, cathode screens, copper end connectors, and copper side plates.

Table 7. Factors affecting cell voltage: the change in cell voltage ΔU_{cell} divided by the change in four important cell factors

Factor	Modified cell *	Standard cell *
ΔU_{cell}/change in current density J, mV m^2 kA^{-1}	450	450
ΔU_{cell}/change in cell temperature t_{cell}, mV/°C	−7.7	−10.1
ΔU_{cell}/change in brine concentration c_{NaCl}, mV L g^{-1}	−0.7	−1.8
ΔU_{cell}/change in cell liquor concentration c_{NaOH}, mV L g^{-1}	0.26	0.6

* OxyTech MDC-55 cell. The modified cell is outfitted with the Modified Diaphragm and expandable anodes, whereas the standard cell is outfitted with the standard asbestos diaphragm and the standard DSA anode. Conditions: anolyte temperature of 93 °C, anolyte NaCl concentration of 250 g/L, and catholyte NaOH concentration of 130 g/L.

temperature, feed brine NaCl concentration, and the cell liquor NaOH concentration also affect cell voltage (Table 7), because they affect the conductivity of the solutions between the electrodes. Table 7 clearly shows that current density is the most important factor.

Excessive brine impurities or other severe operating problems can adversely affect the voltage of the cell.

Power Consumption. The power consumption of a cell, kW h per tonne of Cl_2, may be calculated from the cell voltage by the following equation:

$$\text{Power consumption} = U_{cell} \times 756/\varepsilon$$

where

U_{cell} = cell voltage, V

ε = cell efficiency expressed as a decimal

Optimization. The relationships described in the preceding paragraphs can be used to determine the optimum economical cell operating conditions. The optimizations that must be considered are the following:

1) Higher cell liquor caustic strength and lower steam usage in caustic evaporation versus lower cell efficiency and higher power consumption
2) Lower current density, lower voltage, and lower power consumption versus additional cells and higher capital costs
3) Lower feed brine temperature, thus decreased steam usage for brine heating, versus higher cell voltage, lower efficiency, and higher power consumption
4) High brine pH and reduced acidification costs versus lower chlorine efficiency, higher power consumption, and lower product purity

Each diaphragm cell chlorine plant must determine its own optimum conditions for the most economical operations.

6.2. Diaphragm Cells

Electrolyzers for the production of chlorine and sodium hydroxide, including both diaphragm and membrane cells, are classified as either monopolar or bipolar. The designation does not refer to the electrochemical reactions that take place, which of course require two poles or electrodes for all cells, but to the electrolyzer construction or assembly. There are many more chlor-alkali production facilities with monopolar cells than with bipolar cells.

Bipolar electrolyzers have unit assemblies of the anode of one cell unit directly connected to the cathode of the next cell unit, thus minimizing intercell voltage loss. These units are assembled in series like a filter press, and therefore, the voltage of an electrolyzer is the sum of the individual cell voltages created by the anode of one unit, a diaphragm, and the cathode of the next unit. Bipolar electrolyzers have high voltages and relatively low amperage; therefore, the cost of electrical rectification is lower per unit of production capacity. Bipolar electrolyzers either must be installed in a large number of electrical circuits or be designed with very large individual cell components. Developers have chosen the option for large components.

Dow Chemical was the only early developer to have chlorine production needs large enough to consider the bipolar option [2]. Later, following the development of the DSA anode, PPG Industries and Oronzio De Nora Impianti Elettrochimici designed, and PPG Industries installed Glanor bipolar electrolyzers in a large complex at Lake Charles, Louisiana [12].

The *monopolar* electrolyzer is assembled so that the anodes and cathodes are in parallel. Therefore, the potential difference of all cells in the electrolyzer is the same, and the amperage at any particular current density only depends on the electrode surface area. A monopolar electrolyzer has low voltage and high amperage. The highest amperage rating of the most common modern monopolar cells is ca. 150 kA. Because a monopolar electrolyzer has a voltage of only 3 – 4 V, circuits of up to 200 electrolyzers have been constructed, producing 900 t of chlorine per day.

Diaphragms. The earliest asbestos diaphragms were made of sheets of asbestos paper. Asbestos was chosen because of its good chemical stability and its ion-exchange properties. Asbestos has been relatively inexpensive, since it is a relatively abundant natural material that was already being mined and processed for other industrial purposes, such as insulation.

The deposited asbestos diaphragm developed by Hooker Chemical in 1928 was the most common diaphragm until 1971, when what is now OxyTech Systems developed the Modified Diaphragm. The Modified Diaphragm is a mixture of asbestos and a fibrous fluorocarbon polymer [115]. The polymer stabilizes the asbestos, which in itself lowers cell voltage and also allows for the use of the expandable DSA anode [116]. In its various formulations, the Modified Diaphragm is the most common diaphragm in use today. The Modified Diaphragm still contains a minimum of 75 % asbestos.

Environmental concern over the use of asbestos has increased in recent years. France, Saudi Arabia, and Norway have banned the use of asbestos as a separator in chlorine cells. These nations allowed local chlorine producers several years to install non-asbestos replacement separators in existing diaphragm cells or to replace the cells themselves with membrane cells. Under an amendment to the draft Asbestos Directive adopted by the EU Commission 4 May 1999, continued use of chrysotile asbestos will be allowed in existing diaphragm chlor-alkali plants. This exemption to the general EU ban on marketing and use of chrysotile asbestos will last until the end of a chlorine production unit's lifetime, or until suitable substitute materials for diaphragms are available.

There is also concern in the chlor-alkali-industry for the future supply of asbestos, as most of the North American mines producing the grades of asbestos previously used for chlorine cell diaphragms have closed. At present the most common source of asbestos for chlorine cell diaphragms is Zimbabwe. In addition asbestos disposal costs and regulation have continued to increase. All of theses factors have led the chlor-alkali industry to consider non-asbestos diaphragm technology. Three non-asbestos diaphragm systems are commercially available today.

Chloralp Asbestos-Free Technology. Chloralp, a joint venture of Rhône-Poulenc and La Roche, has developed an asbestos-free diaphragm based on the built-in activation concept. The Chloralp separator is made of two vacuum-deposited layers:

1) The first layer, known as the precathode, is a conductive mat of carbon fibers, containing an electrocatalytic powder to decrease the cathode overpotential
2) The second layer is the diaphragm itself, in which PTFE and inorganic materials have replaced asbestos

The cathodic activation provides energy savings from 50 to 150 mV, depending on the plant operating conditions, and the catalyst content in the mat. Other benefits from the precathode are electrocatalytic destruction of chlorates and lower hydrogen contents in the chlorine after plant shutdowns. The diagraphm is deposited on the precathode by standard vacuum techniques. To control the diaphragm porosity, a pore-forming agent is incorporated in the slurry. Optimized porosity leads to lower cell voltage and higher current efficiency.

The technology is currently evaluated on OxyTech HC-3B electrolyzers at the Chloralp Pont-De-Claix facility. Compared to polymer-modified diaphragms, Chloralp asbestos-free separators provide a 150 mV saving on voltage and a 2% improvement in current efficiency. Combined benefits from the precathode and the diaphragm, as well as an extended lifetime, could lead in the near future to full conversion of the 240 000 t/a Pont-De-Claix plant.

OxyTech Polyramix Diaphragm. OxyTech Systems has developed and commercialized a synthetic non-asbestos diaphragm called the Polyramix diaphragm. The Polyramix fiber is composed of a PTFE [poly(tetraflouroethylene)] fibrid base with zirconium oxide ceramic particles embedded in and on the fiber. The Polyramix diaphragm is vacuum deposited onto the cathode and then baked in an oven to fuse the fibers together. The process is very similar to OxyTech's widely used Modified Diaphragms. The Vulcan Chemicals plant in Geismar Louisiana was fully converted to the use of Polyramix diaphragms in 1993. Most major diaphragm cell chlorine plants have operated from 2 to 40 Polyramix diaphragm cells. The longest life cell with a Polyramix diaphragms has been in operation for over ten years [117].

PPG Industries Tephram Diaphragm. PPG Industries has developed and commercialized the Tephram diaphragm as its entry into the non-asbestos diaphragm market. Major goals of PPG's non-asbestos program were to produce a diaphragm that deposited and operated similarly to asbestos, with longer diaphragm life and lower power consumption. This technology utilized vacuum deposition to produce a base diaphragm composed of PTFE and a topcoat from a slurry of inorganic particulate materials. Dopants are periodically added to the anolyte during cell operation to adjust the diaphragm permeability to maintain or improve cell operation. This diaphragm has been successfully tested of the major monopolar cell technologies, as well as on PPG's Glanor bipolar cells. An active program is in place to operate diaphragms on a trial basis at various producer sites, allowing for site-specific evaluation of the overall economics of the diaphragm.

6.2.1. Dow Cell [2], [118], [119]

The Dow Chemical Company is the largest chlor-alkali producer, accounting for one-third of the U.S. production and one-fifth of the world capacity. Because Dow's production capacity is large and concentrated in a few sites, Dow's cell development followed a different path than other chlor-alkali technology developers. Dow uses its own cell design of the filter press *bipolar* type. Dow has operated filter press cells for

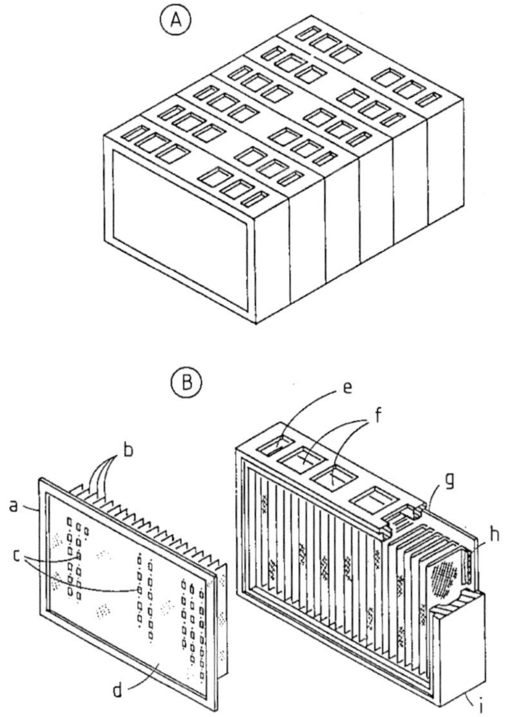

Figure 33. Dow diaphragm cell
A) Six-cell series
B) Internal cell parts: a) Cathode element; b) Cathode pocket elements; c) Copper spring clips; d) Perforated steel backplate; e) Brine inlet; f) Chlorine outlets; g) Copper backplate; h) Titanium backplate; i) Anode element

over 90 years. Dow cell development occurred in several stages, characterized by simple rugged construction and relatively inexpensive materials.

The current cell employs vertical DSA coated titanium anodes, vertical cathodes of woven wire mesh bolted to a perforated steel backplate, and a vacuum-deposited modified asbestos diaphragm. A single bipolar element may have 100 m^2 of both anode and cathode active area. The anode of one element is connected to the cathode of the next by copper spring clips. This connection is immersed in the cell liquor during operation. Figures 33 B and 34 show these internal cell parts.

Dow operates at lower current densities than others in the chlor-alkali industry. The electrolyzers are normally operated with 50 or more cells in one unit or series. One electrical circuit may consist of only two of these electrolyzers. Figure 33 A shows a view of six electrolytic cells.

Treated saturated brine is fed to the anolyte compartment, where it percolates through the diaphragm into the catholyte chamber. The percolation rate is controlled by maintaining a level of anolyte to establish a positive, adjustable hydrostatic head. The optimum rate of brine flow usually results in the decomposition of ca. 50 % of the incoming NaCl, so that the cell liquor is a solution containing 8 – 12 wt % NaOH and 12 – 18 wt % NaCl.

The Dow diaphragm cell, optimized for low current density, consumes less electrical energy per unit of production than the rest of the industry. The cell voltage at these low

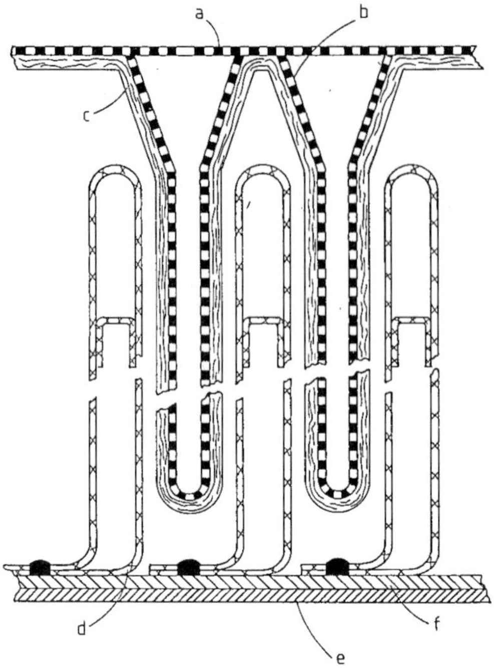

Figure 34. Dow diaphragm cell, section view
a) Perforated steel backplate; b) Cathode pocket; c) Asbestos diaphragm; d) DSA anode; e) Copper backplate; f) Titanium backplate

current densities is only 300 – 400 mV above the decomposition potential of the cell. However, Dow has a larger investment in the electrolyzers, especially anodes.

The electrolyzers are operated at ca. 80 °C, lower than the 95 °C typical of other types of cells. This lower operating temperature allows cell construction with less expensive materials, such as vinyl ester resins and other plastics [118]. Operating data have not been published.

6.2.2. Glanor Electrolyzer [12], [120] – [122]

Glanor *bipolar* electrolyzers are a joint development of PPG Industries and Oronzio De Nora Impianti Elettrochimici S.p.A. The Glanor electrolyzer consists of several bipolar cells clamped between two end electrode assemblies by means of tie rods, thereby forming a filter press type electrolyzer (Fig. 35). The electrolyzer is equipped with DSA titanium anodes. Each electrolyzer normally consists of 11 or 12 cells. A lower number of cells can, however, be assembled in one electrolyzer. The Glanor electrolyzer was especially designed for large chlor-alkali plants.

The current is fed into the electrolyzer by means of anodic and cathodic end elements. The anodic compartment of each cell is connected to an independent brine feed tank by means of flanged connections.

Chlorine gas leaves each cell from the top through the brine feed tank and then passes to the cell room collection system. Hydrogen gas leaves from the top of the

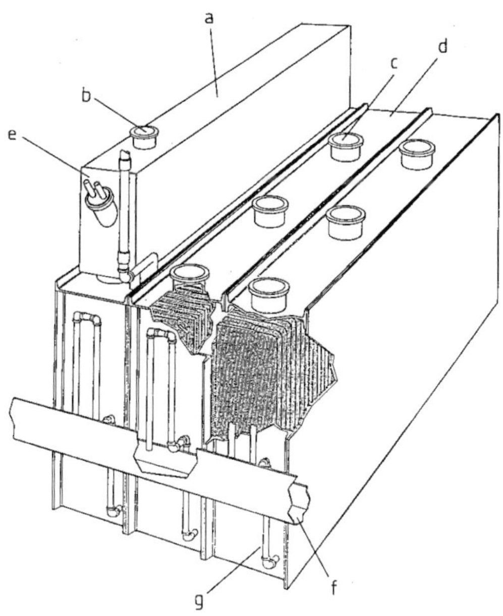

Figure 35. Glanor bipolar electrolyzer
a) Disengaging tank; b) Chlorine outlet; c) Hydrogen outlet; d) Bipolar element; e) Brine inlet; f) Cell liquor trough; g) Cell liquor outlet

Figure 36. Glanor bipolar electrolyzer type V-1144

cathodic compartment of each cell, while the catholyte liquor leaves from the bottom through an adjustable level connection.

The V-1144 electrolyzer (Fig. 36) was the first commercial unit, and eight plants utilize this model. The second generation is the V-1161 electrolyzer, which employs Modified Diaphragms, narrower electrode gaps, lower current density, and DSA anodes to achieve lower power consumption than the V-1144 electrolyzer.

The operating characteristics of the Glanor electrolyzers are shown in Table 8.

Table 8. Glanor bipolar diaphragm electrolyzers: design and operating characteristics

Item	Model V-1144	Model V-1161
Cells per electrolyzer	11	11
Active anode area per cell, m^2	35	49
Electrode gap, mm	11	6
Current load, kA	72	72
Current density ϱ at 72 kA, kA/m^2	2.05	1.47
Cell voltage, V	3.50	3.08
Current efficiency, %	95 – 96	95 – 96
Power consumption (d.c.),		
kW h/t *	2500	2200
Anode gas composition		
(alkaline brine)		
Cl_2, %	97.3 – 98.0	97.0 – 98.0
O_2, %	1.5 – 2.2	1.5 – 2.2
H_2, %	< 0.1	< 0.1
CO_2, %	0.4	0.4
Cell liquor		
NaOH, g/L	135 – 145	135 – 145
$NaClO_3$, %	0.03 – 0.15	0.03 – 0.15
Production per electrolyzer		
Chlorine, t/d **	26.7	26.7
NaOH, t/d	29.8	29.8

* Per short ton of chlorine.
** Short tons.

6.2.3. OxyTech "Hooker" Cells [2], [12], [123], [124]

The first commercialized deposited asbestos diaphragm cell was the Hooker type S-1 monopolar cell, introduced in 1929. The basic design featured vertical graphite anode plates connected to a copper bus bar and a cathode with woven steel wire cloth or perforated steel fingers between the anodes. The cathode held vacuum-deposited asbestos fiber diaphragms that separated the anode and cathode compartments. The cathode fingers did not extend completely across the cell, but left a central circulation space. In the following 40 years, a family of S series cells with similar characteristics evolved, with over 12 000 having been installed in licensed plants.

In 1973, a new H series of *monopolar* cells was introduced. They incorporated the use of DSA anodes, which had been developed and commercialized in the late 1960s. These cells have significant voltage savings over the S series, thus allowing increases in cell capacity without corresponding increases in rectification capacity. The H series also incorporate cathode tubes with both ends open, extending across the cell, as the circulation space requirement was satisfied by the change from solid graphite anodes to the open DSA anodes (Fig. 37).

Table 9 is a summary of operating characteristics and current densities of the H-series cells currently available for license.

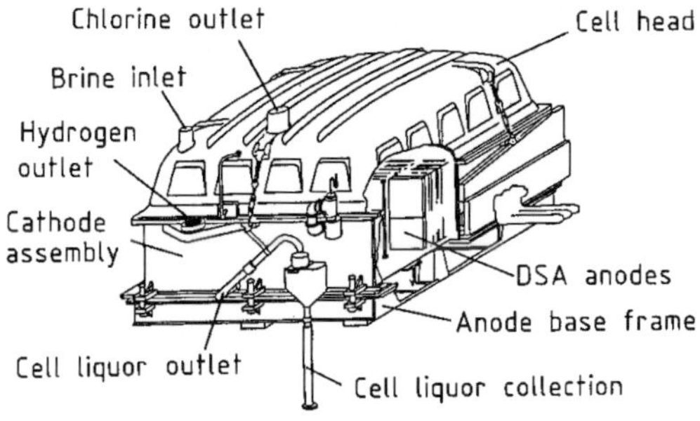

Figure 37. OxyTech type H-4 cell

Table 9. OxyTech Systems Hooker H-series diaphragm cells: design and operating characteristics

	H-2A		H-4	
Operating current, A	80 000		150 000	
Anode area, m^2	36.16		64.52	
$in.^2$	56 050		100 000	
Current density, A/m^2	2212		2325	
$A/in.^2$	1.43		1.50	
Cell voltage, V	3.44		3.44	
Approximate cell dimensions, m	1.87 × 2.66		2.58 × 3.11	
Diaphragm life, days	300 – 500		300 – 500	
Anode life, years	5 – 7	5 – 7		
Operating NaOH concentration, g/L	140	160	140	160
%	11.35	12.89	11.33	12.87
Current Efficiency, %	96.4	94.6	96.6	94.9
Chlorine output,				
metric ton/day	2.45	2.41	4.60	4.52
short ton/day	2.70	2.65	5.07	4.98
Caustic soda output,				
metric ton/day	2.76	2.71	5.19	5.10
short ton/day	3.05	2.99	5.72	5.62

6.2.4. HU Monopolar Cells [124]

The HU type cells were a joint development of Hooker (now OxyTech Systems) and Uhde. The HU-type electrolyzer (Fig. 38) is rectangular, not cubic, and is narrow in the direction of current flow, since anodes are arranged in a single row. The cathode is long and narrow; consequently, the current density is lower through the cathode shell. The long, narrow cathode fabrication lends itself to closer anode – cathode tolerances and spacing. Copper on and around the cathode shell has been eliminated. Another advantage of the long, narrow design is shorter electrolysis current paths through the cell room, resulting in savings in piping and other materials. The HU-type cell incorporates a Modified Diaphragm.

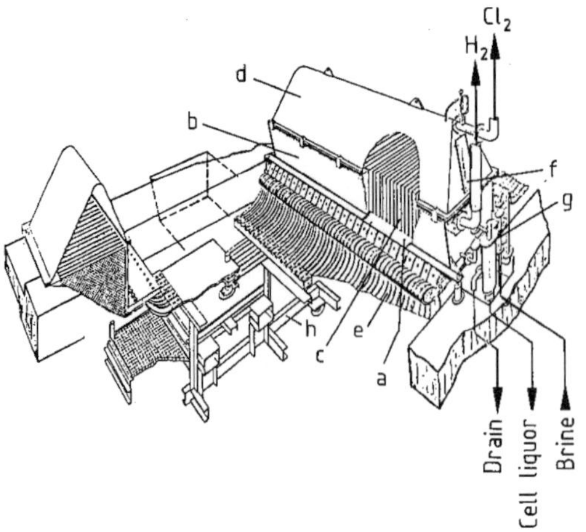

Figure 38. OxyTech/Uhde HU-type cells
a) Cell bottom; b) Cathode; c) Anode; d) Cell cover; e) Bus bars; f) Brine level gauge; g) Brine flow meter; h) Bypass switch

Table 10. HU series diaphragm cells: design and operating characteristics

Item	Cell type						
	HU 24	HU 30	HU 36	HU 42	HU 48	HU 54	HU 60
Number of anodes	24	30	36	42	48	54	60
Anode surface area, m^2	20.6	25.8	31.0	36.1	41.3	46.4	51.6
Load, kA	30 – 45	40 – 60	50 – 70	55 – 85	60 – 95	70 – 105	80 – 120
Cl_2 production, t/d	0.90 – 1.36	1.19 – 1.82	1.49 – 2.12	1.64 – 2.58	1.79 – 2.88	2.09 – 3.18	2.39 – 3.64
NaOH (100 %) production, t/d	1.01 – 1.54	1.35 – 2.05	1.68 – 2.39	1.85 – 2.91	2.02 – 3.25	2.36 – 3.59	2.69 – 4.10
H_2 production, kg/d	25 – 39	34 – 52	42 – 60	47 – 73	51 – 82	59 – 91	68 – 103
Cell length, m	2.1	2.6	3.0	3.5	3.9	4.4	4.8
Distance, cell-to-cell, m	1.5	1.5	1.5	1.5	1.5	1.5	1.5

A further novelty of the HU cell system is the design and arrangement of the bypass switch. The HU switch is installed underneath, not next to, the circuit of cells. This is accomplished by raising the cells from the floor, similar to mercury cells, creating a second operating floor. The interconnecting bus bars are flexible and are distributed over the entire length of the cell. The HU cell design incorporates a bus bar for each individual anode. This, as well as the elevation of the cell from the floor below, which allows access, enables connection of facilities for monitoring the current flowing through each anode. During operation of the bypass switch, connection is made for each individual anode, and no additional contact bus bars are required.

The HU-type cells are offered to cover 30 – 150 kA. All of the different cell types are equipped with cathodes and anodes of identical height and width. The only basic difference between the various cell models is the number of elements and consequently the length of the cell (Table 10). Cell voltage and power consumption per tonne of chlorine, identical for all cell types, are shown in Table 11 for the specific current loads of 1.5 and 2.3 kA/m^2.

Table 11. HU series diaphragm cells: specific load, cell voltage, and power consumption

	Specific load, kA/m²	
	1.5	2.3
Cell voltage, V	3.12	3.41
Power consumption (d.c., average), kW h/t*	2500	2700

* Per tonne of chlorine.

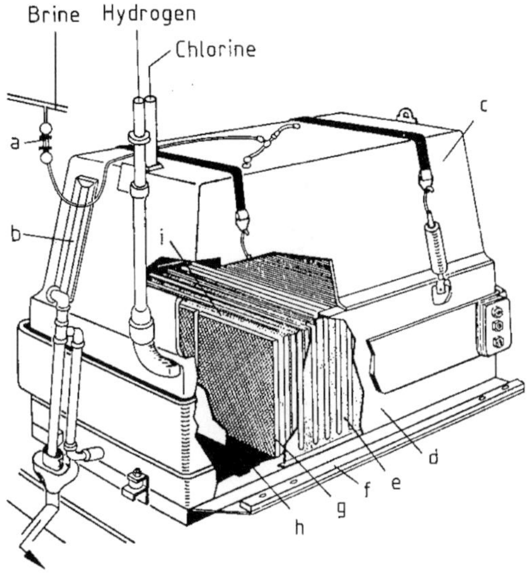

Figure 39. OxyTech Systems MDC cells
a) Brine feed rotometer; b) Head sight glass; c) Cell head; d) Cathode assembly; e) Tube sheet; f) Grid plate; g) Cathode tube; h) Grid protector; i) DSA expandable anode

6.2.5. OxyTech MDC Cells [12], [125]

OxyTech Systems manufactures and licenses the MDC series of *monopolar* diaphragm cells (Fig. 39). The MDC cells feature woven steel wire cathode screen tubes open at both ends, which are welded into thick steel tube sheets at each end. The tubes, tube sheets, and the outer steel cathode shell form the catholyte chamber of the cell (Fig. 40). Copper is bonded, rather than welded, to the rectangular cathode shell on the two long sides parallel to the tube sheets. Copper connectors attached at the ends of the bonded copper side plates complete the encompassing of the cathode with copper. Anodes are connected to a copper patented cell base, which is protected from the anolyte by a rubber cover or a titanium base cover (TIBAC) [126]. Orientation of the cathode tubes is parallel to the cell circuit, the opposite of a Hooker-type cell. This arrangement accommodates thermal expansion of the cell and circuit without changing the anode-to-cathode alignment.

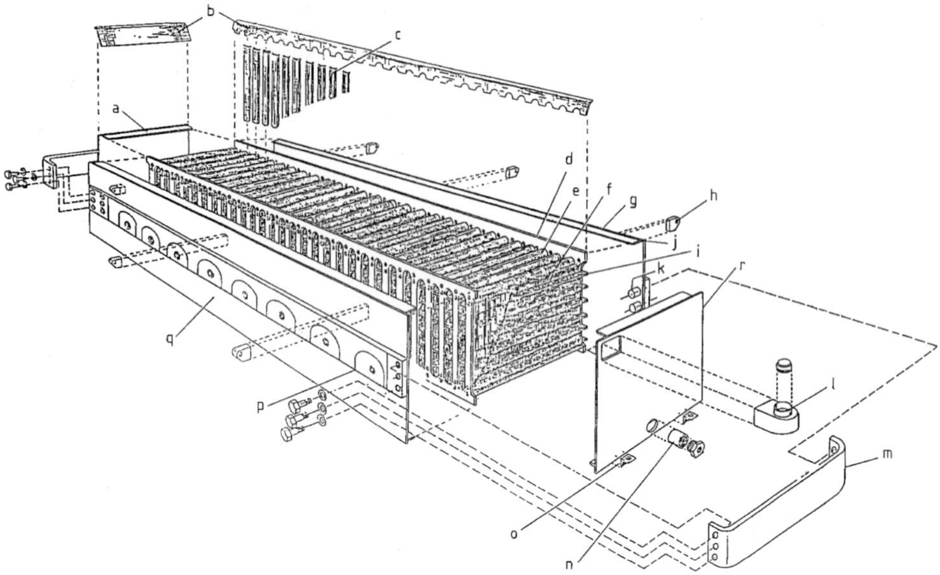

Figure 40. Exploded view of an OxyTech MDC-55 cathode
a) End plate; b) Rim screen; c) Side screens; d) Tube sheet; e) Full cathode tube; f) Half-cathode or end tube; g) Side plate; h) Lifting lug; i) Punched and coined stiffener strap; j) Bosses; k) End plate, operating aisle end; l) Hydrogen outlet; m) Connector bar; n) Caustic outlet; o) Clip angles; p) Grid bar, connector side; q) Side plate

The combination of the Modified Diaphragm and expandable DSA anodes reduces power consumption by 10 – 15 % from that of regular asbestos diaphragms and standard, fixed DSA anodes [127]. Table 12 presents performance data for the two most common MDC cell sizes [125]. The OxyTech MDC-29 is shown in Fig. 41. The licensed chlorine capacity of OxyTech cells now exceeds 20 000 t/d.

6.3. Operation

The process description in this section is intended to provide an overview of typical diaphragm cell process areas. A general block diagram for a diaphragm cell facility is shown in Figure 12. Included on the drawing are many process areas that may be optional, depending on the design of the plant and its end products. The operation of a cell room may be broken down into six areas: the two incoming systems, brine and electrical; the cells; and the three outgoing systems, chlorine, hydrogen, and cell liquor. Some of these are essentially the same for all three chlor-alkali processes and are described in Chapter 4 — the brine system (general), the electrical system, and the hydrogen system. The treatment of the chlorine is the subject of Chapter 11. Only aspects that are reasonably specific to the diaphragm cell process are described in this section.

Table 12. OxyTech Systems MDC cells: operating capacities and characteristics

Item	Model number and operating range, kA			
	MDC-29		MDC-55	
	35 to	80	75 to	150
Chlorine capacity,				
metric ton/day	1.05	2.41	2.33	4.53
short ton/day	1.16	2.66	2.48	5.00
Caustic capacity,				
metric ton/day	1.21	2.76	2.59	5.18
short ton/day	1.33	3.04	2.85	5.70
Hydrogen capacity,				
m^3/day	335	765	720	1435
cubic feet/day	11 830	27 010	25 420	50 670
Current density,				
kA/m^2	1.21	2.76	1.37	2.74
$A/in.^2$	0.78	1.78	0.88	1.76
Cell voltage, V [a]				
steel cathode	2.90	3.62	3.00	3.62
activated cathode	2.80	3.51	2.90	3.51
Power consumed (d.c., steel cathode) [b],				
kW h/t	2310	2876	2390	2870
kW h/short ton	2100	2610	2175	2610
Power consumed (d.c., activated cathode) [b],				
kW h/t	2230	2786	2310	2780
kW h/short ton	2025	2530	2100	2530
Diaphragm life, years	1–2	0.5–1.0	1–2	0.5–1.0
Anode life, years	8–10	5–8	8–10	5–8
Cathode life, years	10–15	10–15	10–15	10–15
Distance between cells [c],				
m	1.60		2.13	
inches	63		84	

[a] Cell voltage includes loss in intercell bus.
[b] Power consumed per ton (metric or short) of chlorine produced.
[c] Distance centerline-to-centerline and side-by-side with bus connecting.

Figure 41. OxyTech MDC-29 with the author (1971)

Table 13. Typical brine feed specifications for diaphragm cells

Parameter	Specification
NaCl	$\geq$320 g/L
pH	2.5 – 3.5
Hardness (Ca^{2+} + Mg^{2+})	< 5 ppm
Magnesium	< 0.4 ppm
Sodium sulfate (Na_2SO_4)	< 5 g/L
Organics	< 1 ppm
Manganese	< 0.01 ppm
Barium	< 0.01 ppm
Nickel	< 0.1 ppm
Iron	< 0.5 ppm
Silicon	< 15 ppm
Cobalt	< 0.02 ppm
Mercury	< 1 ppm
Phosphate	< 1 ppm

6.3.1. Brine System

Most commonly, diaphragm cells are supplied with well brine on a once-through basis. The treated well brine flows to the treated brine storage tanks, which usually have 12-h capacity. From there the brine is fed to the cell room. The flow to each individual electrolyzer is controlled by a rotameter. If the flow of brine to the cells is suddenly disrupted by failure of the brine feed pump, the rectifiers automatically shut down since an inadequate supply of brine to the cells is potentially unsafe. The specifications for brine for diaphragm cells are given in Table 13.

A brine recovery lagoon is usually available to handle any major upsets in the brine system. Brine sludges or out-of-spec brine can be sent to the lagoon. Supernatant clear brine can be recovered from the lagoon.

In most cases, operation with acidic brine is preferred because of the reduced amount of side-reaction products in the chlorine and the cell liquor.

Figure 42. Cell room: bipolar PPG Industries/De Nora Glanor cells

Figure 43. Cell room: monopolar OxyTech H-4 cells

6.3.2. Cell Room

Typical cell rooms are shown in Figures 42 (bipolar cells) and 43 (monopolar cells). A cell in normal operation requires little attention. The critical requirement is that the *brine flow rate* is sufficient to maintain an anolyte level above the cathode.

Under no circumstances should a cell be operated with an inadequate or excessive anolyte level. Operation with the anolyte level not visible in a sight glass is unsafe. At least one operator should be in the cell room at all times. The cell room operator should inspect the anolyte level and brine flow to each cell at least once per hour. Any change in the anolyte level or brine flow rate should be investigated.

As the cell ages, the diaphragm will undergo changes in porosity because of the following:

1) Electrolysis effect
2) Brine impurities
3) Upsets in operation
4) Gradual wear of the diaphragm

A change in porosity may necessitate a change in brine flow rate. If the increase in porosity is severe, the cell may be replaced or doped with an asbestos slurry or inorganic salts.

Impurities in the brine often lead to decreased porosity. Decreased porosity can be offset to some extent by increasing the anolyte level and, if necessary, by lowering the catholyte level.

A cell operated with the anolyte level at the maximum value and the lowest catholyte level is called a sleeper. To gain additional diaphragm life after a cell has entered the sleeper position, the brine flow rate must be decreased below normal. This is not normally a recommended practice because current efficiencies of these cells are usually low.

For safe operation of diaphragm cells, the *header pressures* must be maintained at the proper values.

The chlorine header should be maintained at positive pressure to permit detection and correction of any chlorine piping leaks. The hydrogen header is also maintained at a positive pressure to avoid pulling air into the hydrogen, creating a potentially explosive mixture. The brine header pressure should be maintained to give the desired caustic concentration in the cell liquor. Normal practice is to adjust individual brine feed valves so that each cell receives the correct brine flow rate.

Load changes must be smooth to avoid fluctuations in the header pressures and detrimental effects on the diaphragms.

The brine feed rate to each cell should be increased to the new rate before circuit amperage is increased. The brine feed rate to each cell should be decreased immediately after amperage is decreased. During any period of operation when brine flow rates are being changed, extra attention should be given to the anolyte levels of the cells. Adjustment of the brine feed temperature may also be necessary when a load change occurs.

6.3.3. Diaphragm Aging

Of all the cell components, the diaphragm usually has the shortest life. The ability of a diaphragm to resist the back migration of hydroxide slowly becomes impaired with service life. The performance of the diaphragm deteriorates for the following reasons:

1) Chemical attack
2) Brine impurities
3) Unsteady operating conditions

The major reason for the deterioration is *chemical attack* on the asbestos by the alkaline catholyte and acidic anolyte. The rate of chemical attack can be minimized and diaphragm life maximized by careful operation of the cell. The most important situations to avoid are high concentrations of brine impurities and unsteady operating conditions. High *brine impurities* cause plugging of the diaphragm with insoluble hydroxides, which reduce the diaphragm's separation ability. The most common harmful impurities are calcium, magnesium, iron, nickel, silicates, aluminum, manganese, and barium.

Unsteady operation, such as electrical load changes, cell liquor strength changes, changes in brine concentration or pH, gas-pressure fluctuations, and shutdowns, change the pH of the various regions of the diaphragm, thus accelerating chemical attack on the asbestos. Diaphragm cell plant operators should strive to minimize these changes.

The real importance of the equations in Section 6.1 is as an aid in deciding when the diaphragms should be replaced.

6.3.4. Treatment of the Products

Chlorine. See Chapter 11.

Hydrogen. See Chapter 4.

Sodium Hydroxide Solution. The hydroxide produced at the cathode is associated with sodium ions and water to form a 10 – 12 wt % sodium hydroxide solution leaving the electrolytic cell. This cell liquor also contains 18 wt % unreacted sodium chloride.

Most large modern chlor-alkali plants have or will soon have an associated cogeneration power plant. In these facilities, the caustic evaporators are an important use for the byproduct steam.

Modern diaphragm cell plants use triple-effect evaporators and, in many cases, quadruple-effect evaporators.

Caustic Soda Evaporation. A flow diagram for a typical triple-effect caustic soda evaporator is shown in Figure 44. The evaporator is of the backward-feed design and concentrates 10 – 11.3 wt % NaOH cell liquor to 50 wt % NaOH. Liquor flows from

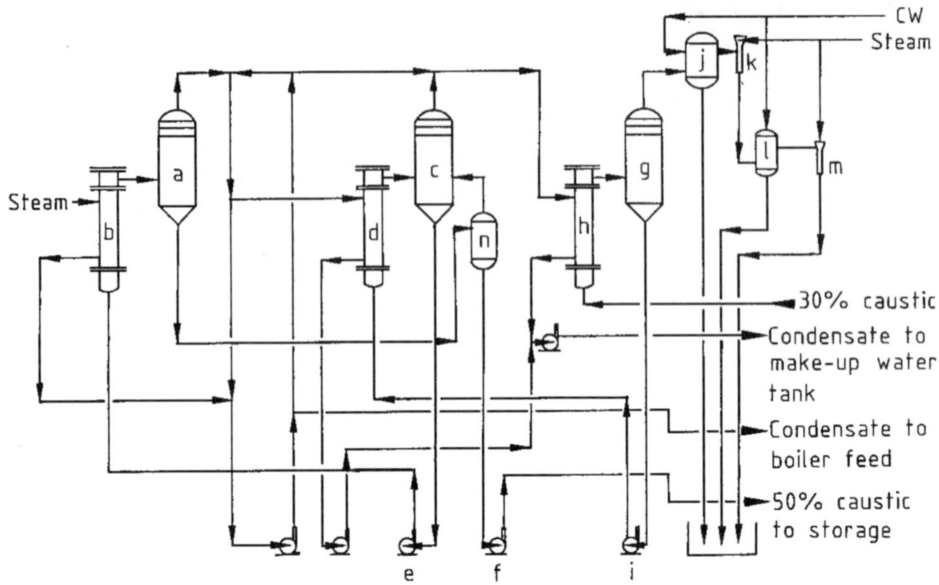

Figure 44. Process flow diagram: triple-effect caustic evaporator
a) First-effect vapor body; b) First-effect heat exchanger; c) Second-effect vapor body; d) Second-effect heat exchanger; e) Second-effect forwarding pumps; f) 50 % caustic transfer pumps; g) Third-effect vapor body; h) Third-effect heat exchanger; i) Third-effect forwarding pump; j) Barometric condenser; k) First-stage ejectors; l) Intercondenser; m) Second-stage ejector; n) Liquor flash tank

the third to the second to the first effect and from the first effect to the liquor flash tank. A cyclone is used for each effect to utilize the pressure drop across the circulating pump to clarify the transfer liquor.

The salt precipitated in the liquor flash tank is isolated from the rest of the salt precipitated in the evaporator and used as seed crystals in the cooling system to help diminish coil scaling and supersaturation of the product liquor with sodium chloride. The sodium chloride and triple salt ($NaOH-NaCl-Na_2SO_4$) precipitated in the liquor flash tank and cooling system is removed from the cooled product liquor with centrifuges. The salt precipitated in the three effects flows countercurrent to the liquor flow so that all of the salt is discharged from the last effect, the effect that has the coldest liquor and the lowest caustic soda concentration.

Two-stage steam-jet air ejectors with a common intercondenser are used to maintain vacuum in the evaporator. In the caustic cooling system, agitated tanks are used to cool the slurry discharged from the liquor flash tank. The slurry flows from the cooling system to the centrifuge feed tank from where it is pumped into centrifuges. Salt discharged from the centrifuges drops into the evaporator feed tank, where it is dissolved in cell liquor. The 50 wt % NaOH concentrate liquor, which flows by gravity to the pressure filter feed tank, contains ca. 1.0 – 1.5 wt % dissolved NaCl and ca. 0.1 wt % crystalline NaCl.

The liquor is pumped from the pressure filter feed tank into the pressure leaf filters, where the remaining traces of salt crystals are removed. The product caustic flows by gravity to the filtered product tank and then is pumped to storage. Salt removed in the pressure filters is reslurried with cell liquor and pumped to the evaporator feed tank via the filter backwash pump.

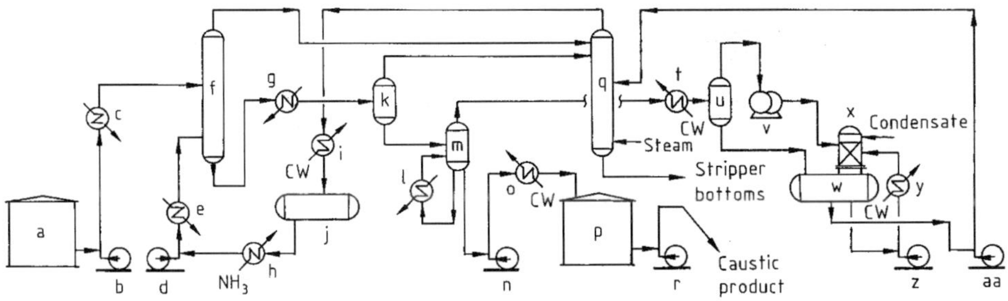

Figure 45. Caustic purification system
a) 50 % caustic feed tank; b) 50 % caustic feed pumps; c) Caustic feed preheater; d) Ammonia feed pumps; e) Ammonia feed preheater; f) Extractor; g) Trim heater; h) Ammonia subcooler; i) Stripper condenser; j) Anhydrous ammonia storage tank; k) Primary flash tank; l) Evaporator reboiler; m) Evaporator; n) Caustic product transfer pumps; o) Purified caustic product cooler; p) Purified caustic storage tank; q) Ammonia stripper; r) Purified caustic transfer pumps; t) Overheads condenser; u) Evaporator; v) Evaporator vacuum pump; w) Aqueous storage ammonia tank; x) Ammonia scrubber; y) Scrubber condenser; z) Ammonia recirculating pump; aa) Ammonia recycle pump

The salt discharged from the centrifuges drops into the leaching tank, where it is reslurried with condensate and recycled brine from the Glauber's salt ($Na_2SO_4 \cdot 10\,H_2O$) crystallizer. Concentrate from the pusher centrifuges flows by gravity into the evaporator feed tank.

The product salt is discharged from a cyclone into the salt reslurry tank. The overflow from the cyclone is returned to the leaching tank. The product salt is diluted with brine and pumped to the resaturator tank.

Brine containing the dissolved sodium sulfate is separated from the salt crystals in a cyclone. The underflow returns to the leaching tank. The overflow is collected in the feed tank for the Glauber's salt crystallizer. Sodium sulfate is crystallized from the liquor in a continuous vacuum cooled crystallizer.

Mother liquor removed from the crystallizer is pumped under level control to the brine tank. Slurry discharged from the crystallizer is thickened to ca. 50 wt %. The liquor from the thickener is collected in the brine tank and pumped back to the leaching tank. The thickened slurry is redissolved in the Glauber's salt dissolving tank and pumped to a waste treatment system.

Caustic Purification [128]. Diaphragm-cell chlor-alkali producers requiring higher purity caustic than that produced by the diaphragm process can use caustic purification or DH process (Fig. 45). *Salt removal* in the purification unit is effected by contacting the 50 wt% caustic with anhydrous liquid ammonia under pressure sufficiently high to maintain all materials in the liquid state.

The liquid ammonia absorbs salt, chlorate, carbonate, water, and some caustic. It is then stripped, concentrated, and returned to the extraction process. The concentrated caustic leaving the extractor is stripped free of ammonia, which is recovered, concentrated, and recirculated. Typical purities before and after caustic purification are shown in Table 20. This process is offered for license by PPG Industries and OxyTech Systems.

In addition, producers and users of diaphragm-cell caustic may wish to *reduce metal impurities* by utilizing the porous cathode cell process (PPG Industries) [129]. The process consists of an electrolysis cell with porous nonmetallic cathodes. The caustic soda (50 wt %) is freed from iron, nickel, lead, and copper, which are deposited on the cathode. The cell must be regenerated periodically with water and hydrochloric acid. Typical feed and product analyses based on anhydrous NaOH are

Metal	Content, ppm	
	Feed	Product
Iron	10.0	2.0
Nickel	3.0	0.2
Lead	4.0	0.4
Copper	0.2	0.1

6.3.5. Measurement

Recorded data is an important tool for determining the operating condition of the plant and diagnosing problems.

The following should be recorded continuously or hourly:

ampere load on each circuit
voltage for each circuit
chlorine header pressure
hydrogen header pressure
brine header pressure or flow rate
brine temperature
brine pH
cell liquor temperature

Samples of brine should be taken every 4 h and combined into a daily composite. In addition, samples of cell liquor should be taken from each cell string, the sodium hydroxide content analyzed, and the temperature taken every 4 h. A daily composite should be made and samples should be analyzed by the laboratory for the following:

NaOH content
NaCl content
salt : caustic ratio
$NaClO_3$ content
NaOCl content
Fe content
average temperature
specific gravity at 25 °C

Chlorine gas from each cell circuit should be analyzed for chlorine and hydrogen content at least twice each 8-h shift. Each day a complete analysis of the chlorine header gas should be made. Additions or extensions to list may be dictated by plant operation.

Each plant must develop a procedure for taking *individual cell data* so that individual cells may be scheduled for renewal. The following is a minimum schedule:

weekly:	voltage and cell liquor composition (NaOH, NaCl, $NaClO_3$)
monthly:	chlorine composition (Cl_2, O_2, H_2, CO_2, N_2)

7. Membrane Process

In the membrane process, the anolyte and catholyte are separated by a cation-exchange membrane that selectively transmits sodium ions but supresses the migration of hydroxyl ions from the catholyte into the anolyte. A strong caustic soda solution with a very low sodium chloride content can be obtained as the catholyte efflux. The advantages of the membrane process are its energy efficiency and its ability to produce caustic soda of high quality, with almost no impact on the environment. Depending on the particular design, membrane sizes from 0.2 to 5 m^2. The production capacity of an electrolyzer can be up to 90 t/d NaOH (100 %).

The process was started in the early 1970s with development of the perfluorosulfonate membrane Nafion by DuPont [130]. In 1975, a perfluorocarboxylate membrane capable of producing 35 wt % caustic soda became available, from Asahi Glass in Japan [131]. In 1978 the first two-layer membrane was developed, with low electrical resistance and high current efficiency [132].

The industrial success of the membrane process started in Japan, where the abolition of the mercury process on environmental grounds had been promoted by the government. Today the membrane process is the state of the art process for producing chlorine and caustic soda or potassium hydroxide.

The production capacity of chlor-alkali plants using the membrane process reached about 21 % of total world production capacity in 1995 and is predicted to increase to about 28 % in 2001 [133].

7.1. Principles

In a membrane cell a cation-exchange membrane separates the anolyte and catholyte, as shown in Figure 46. Saturated brine is fed into the anode compartment, where chlorine gas is evolved at the anode:

$$2\,Cl^- \rightarrow Cl_2 + 2\,e^-$$

The anolyte is discharged from the cell. The electric field causes hydrated sodium ions to migrate through the membrane into the catholyte. In the cathode compartment, hydrogen is evolved at the cathode, leaving hydroxyl ions, which together with permeating sodium ions constitute the caustic soda:

$$2\,H_2O + 2\,e^- \rightarrow H_2 + 2\,OH^-$$

$$Na^+ + OH^- \rightleftarrows NaOH$$

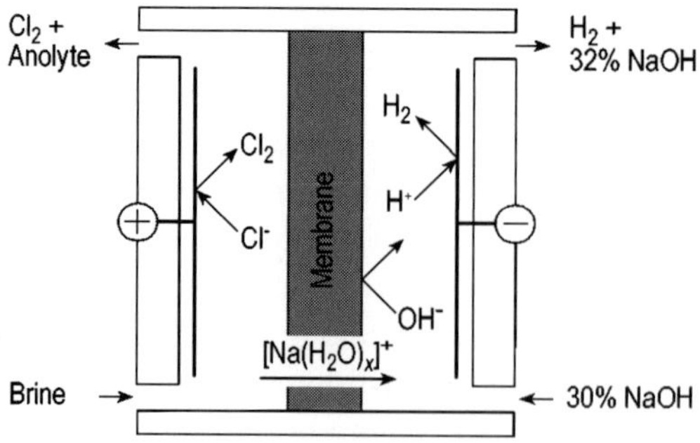

Figure 46. Principle of the membrane cell

Liquid and gaseous phases anolyte/Cl_2 and catholyte/H_2 can be separated either in the cell compartment or downstream of the cell outlet. The chlorine-saturated anolyte is then treated in a dechlorination unit to recover the dissolved chlorine.

Membrane. *Structure.* The membrane is exposed to chlorine and anolyte on one side and strong caustic solution on the other side at high temperature (90 °C). Only ion-exchange membranes made of perfluoropolymer can withstand such severe conditions. The ion-exchange groups of the original polymers are in the fluorosulfonate form, $-SO_3F$, or the carboxylate form, –COOR.

Fluorosulfonate form

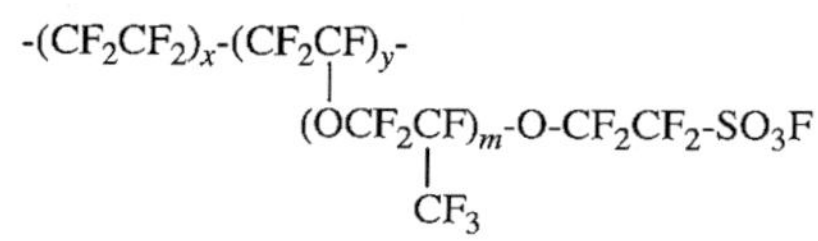

$m = 0 - 1$

Carboxylate form

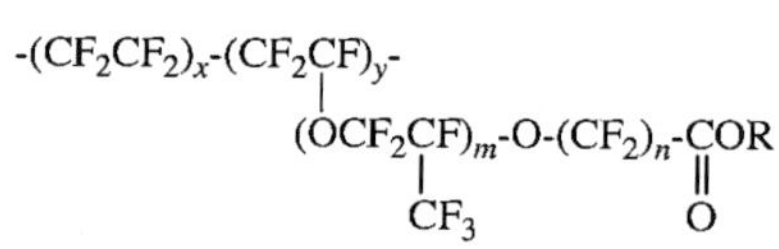

$m = 0 - 1,\ n = 1 - 5,\ R = \text{alkyl}$

Preparation of the Membranes. Modern cation exchange membranes consist of three layers, the carboxylic layer on the cathode side, the sulfonate layer on the anode side and a reinforcement layer of fabric in between.

The active layers are copolymers of tetrafluoroethylene and perfluorovinylethers, that contain esters or other acid precursor groups [308].

The vinyl ethers are prepared by reaction of hexafluoropropylene oxide and methyl-3-fluorocarbonyl perfluoropropionate, followed by pyrolysis to give

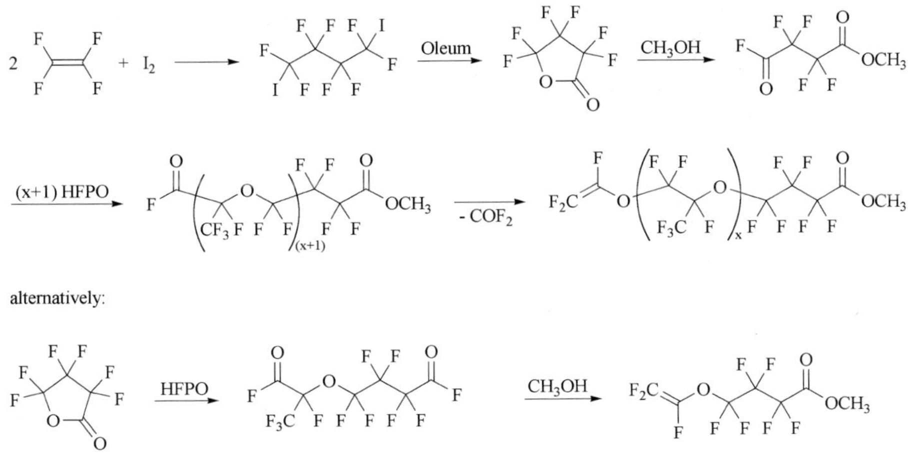

Figure 47. Preparation of a carboxylated perfluorovinyl ether, type Flemion

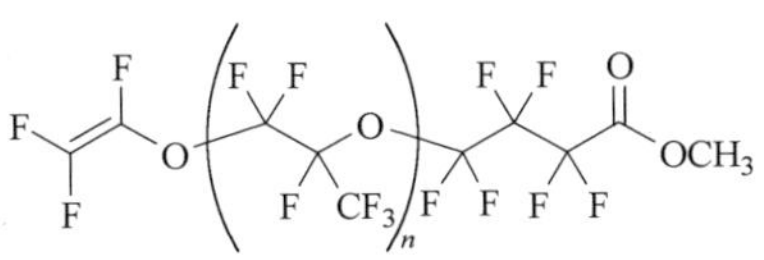

Copolymerisation with tetrafluoroethylene followed by saponification produces the polymer with terminal carboxylic acid groups.

Condensation of hexafluoropropylene with fluorosulfonylperfluoroalkyl carbonyl (prepared by electrochemical fluorination of the respective aliphatic sulfone), gives perfluoroether fluorosulfonyl acyl fluorides which are subsequently converted to the vinyl ether.

Copolymerisation with tetrafluoroethylene gives the polymers with fluorosulfonyl side chaines.

A practical route to a carboxylated perfluorovinyl ether is shown in [309]:

A cyclic lactone is formed by the reaction of 1,4 diiodo-perfluorobutane with oleum (see Fig. 47). The addition of methanol to the lactone gives selectively 3-methoxycarbonyl perfluoro-propionyl fluoride, to which hexafluoropropylene oxide is added. After copolymerization with tetrafluoro ethylene, the acid fluoride group is converted to a perfluorovinyl group by pyrolysis.

The preparation of a sulfonate ion exchange membrane, type "Nafion" from DuPont, is described in [310]:

Tetrafluoroethylene is reacted with sulfur trioxide to give a rearranged sultone. Hexafluoropropylene is oxidized with oxygen to give hexafluoropropylene oxide (HFPO).

The sultone and the HFPO are reacted to give the vinyl ether. Copolymerization of the vinylether with tetrafluoro ethylene gives the perfluorinated polymer (see Fig. 48).

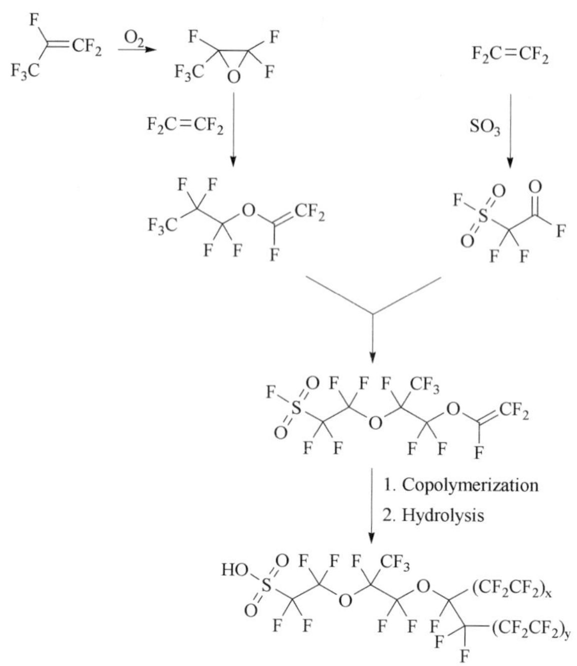

Figure 48. Preparation of a Sulfonated Cation Exchange Membrane, Type "Nafion"

The two copolymers are extruded seperately as films and are afterwards laminated together with the fabric made of polytetrafluoroethylene to give the membrane.

Finally the laminate is hydrolysed in caustic solution to convert the fluorosulfonyl side chains to sulfonate group side chains. See Fig. 49 [311].

As a final step, the surfaces of the membranes can be coated with hydrophilic inorganic porous layers, so as to enhance the release of the gas bubbles generated in the electrolysis cells.

The first membranes that showed significant potential for use in the chlor-alkali process were made of a perfluorosulfonate layer. These proved to be durable in chlor-alkali cells but were relatively inefficient. The perfluorocarboxylate polymers, with a lower water content, showed higher selectivity but led to higher electrical resistance and high electrical power consumption [134]. Combining the advantages of high current efficiency and low electrical resistance a composite membrane (Fig. 50) was developed with a layer containing SO_3^- groups on the anode side and a layer containing COO^- groups on the cathode side [132], [135]. The perfluorosulfonate layer is thicker than the perfluorocarboxylate one and is the major constituent of the membrane [134].

Flux through the Membrane. The total flux through the membrane can be divided in to three parts [134]:

1) Migration due to electric field
2) Convection
3) Diffusion due to chemical gradients

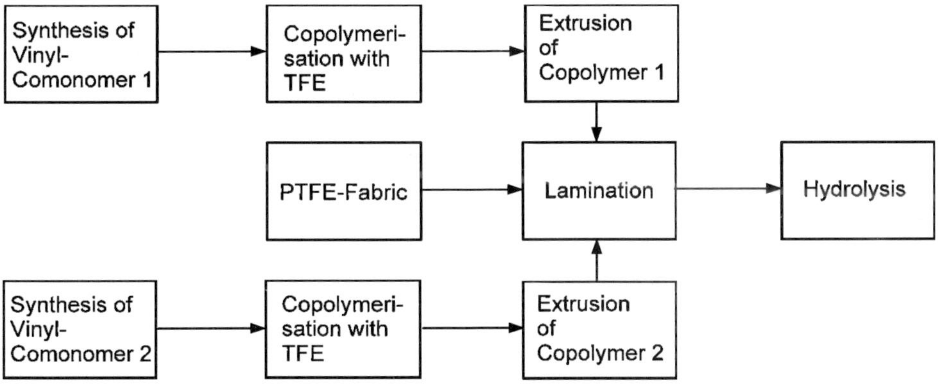

Figure 49. Preparation of the membrane [311]

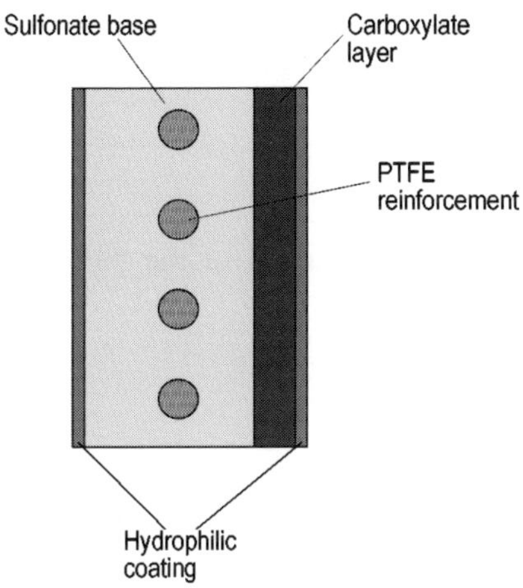

Figure 50. Membrane structure

Migration is the flux of ions through the membrane, driven by electric field. This includes the desired transfer of sodium ions to the cathode compartment and the undesired transfer of hydroxyl ions to the anode compartment. The capacity for selective separation of the cation exchange membrane is determined by its repulsive force for hydroxyl ions. This effect determines the current efficiency.

The backmigration of hydroxyl ions increases the formation of oxygen, hypochlorite, and chlorate in the anode compartment and causes a loss of current efficiency of 3 – 7 % in caustic soda production. The evolution of oxygen gas can be depressed by selecting an anode coating with suitable characteristics (see Section 8.1) or by decreasing the pH in the anode compartment by acidifying the inlet brine (Fig. 51).

Convection and diffusion determine the flow of uncharged compounds and ions through the membrane. Chloride anions in the catholyte are excluded by the cation-exchange membrane and repelled by the electric field, so that the transfer rate of chloride anions from the anolyte is extremly low. As a result, a caustic soda solution of

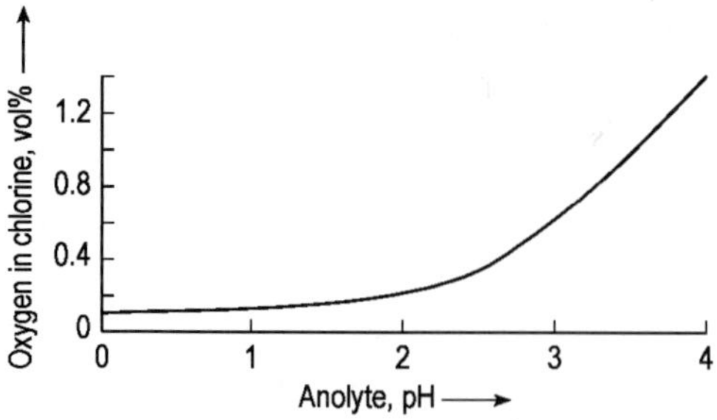

Figure 51. Oxygen content in chlorine [136]

about 32 – 35 wt % with a salt content of less than 20 ppm can be obtained. The water transport through the membrane is about 3.5 to 4.5 moles of water per mole of sodium ions, and can be regarded as the hydration sphere of the migrating sodium ions. Water flux increases with decreasing anolyte concentration [137].

Migration, convection and diffusion influence each other, and the resulting flux depends on membrane type, current density, temperature, and composition of anolyte and catholyte.

Cell Voltage. The cell voltage of a membrane cell is composed of the following terms:

1) Decomposition voltage
2) Membrane potential between anolyte and catholyte
3) Electrode overpotentials for chlorine and hydrogen
4) Ohmic drop in the membrane
5) Ohmic drop in the electrolytes
6) Ohmic drop in electrodes and conductors

Term 1. The decomposition voltage of the chlor-alkali process is about 2.20 V, depending on temperature, concentration, and pressure.

Term 2. This term describes the overpotentials at the surfaces of the membrane. Under standardized operating conditions (3 kA/m^2, 90 °C, 32 wt % caustic solution), the membrane potential is approximately 0.08 V.

Term 3. Titanium anodes coated with oxides of Ir, Ru or Pt are generally used in membrane cells and lead to a chlorine overvoltage of approximately 0.05 V at 3 kA/m^2. Hydrogen overpotentials of about 0.1 V at 3.0 kA/m^2 are attained with activated cathodes. Mainly nickel substrates are coated by painting and thermal treatment or galvanic deposition [138], [139]. Coating materials include Ni, Co, Ru and others (see Section 8.2).

Term 4. The ohmic drop of advanced commercial membranes under standardized operating conditions is about 0.25 – 0.30 V at 3 kA/m^2.

Term 5. To minimize the ohmic drop of an electrolyte, the gaps between the membrane and the electrodes are minimized in membrane cells. However, if the gap is very small, a rise in voltage is observed due to the entrapment of gas bubbles between the electrodes and the hydrophobic fluoropolymer membrane.

Term 6. The voltage losses in the electrolysis cell that occur due to unfavorable current paths along the metallic structure are reduced by an appropriate design. In modern chlor-alkali electrolysis cells the typical ohmic drop is 20 – 40 mV at 3 kA/m^2.

Current Efficiency. The current efficiency (CE) for caustic soda can be obtained either by directly measuring the quantity of caustic soda produced or by an anodic balance, i.e., the compositions of the anode gas and the anolyte with the following equation:

$$CE(\%,NaOH) = 100 - \eta O_2 - \eta ClO_3 - \eta ClO + \eta NaOH + \eta Na_2CO_3$$

where ηO_2, ηClO, and ηClO_3 represent the loss of current efficiency due to the generation of oxygen, hypochlorite, and chlorate, while $\eta NaOH$ and ηNa_2CO_3 take into account NaOH and Na_2CO_3 introduced in the feed brine.

The current efficiency is mainly dependent on membrane performance. On the cathode side the membrane is in contact with a concentrated NaOH solution, while the anode side has a pH of approximately 2 – 4. This leads to a pH gradient across the membrane cross section. The solubility of impurities, which are always present in the pure brine, depends on the pH. Therefore, depending on the type of impurities and on the pH, precipitation inside the membrane can take place. This leads to mechanical destruction of the membrane, which has a irreversible effect on current efficiency. In addition, the cell voltage rises due to the crystals formed inside the membrane.

7.2. Process Specific Aspects

The performance of a membrane cell depends on the following operating conditions:

1) Concentration of anolyte and catholyte
2) Current density
3) Temperature
4) Brine impurities

The optimum caustic strength depends on the composition of the membrane polymer. To achieve stable operation with high current efficiency, fluctuations in operating conditions or upsets must be avoided. Fluctuations in the caustic strength beyond the optimum range influence both the current efficiency and the cell voltage, as shown in Figure 52 [132]. Dilution of the anolyte caused by an upset in the brine feed also decreases the current efficiency. The sensitivity of membrane performance to operating conditions is attributed to changes in the water content of the membrane.

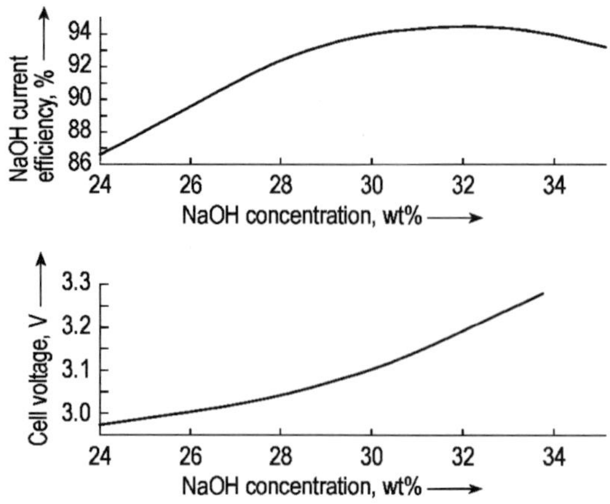

Figure 52. Dependence of cell voltage and current efficiency on NaOH concentration [140]

7.2.1. Brine Purification

The introduction of membrane technology into chlor-alkali electrolysis has dramatically increased the demands on brine purity [141]. The lifetime of chlor-alkali membrane cells is determined by the operating conditions and the quality and purity of the feed into the electrolyzers. Good long-term performance of the cells may be obtained if brine impurities are kept within the limits recommended in Table 14.

A major source of performance decline is the accumulation of solid material in the membrane [142]. Specific impurity levels are dependent on membrane design, cell design, operating conditions, the impurity itself and other impurities present. The prerequisite for long membrane life is to maintain low levels of, for example, Ca^{2+}, Mg^{2+}, Sr^{2+}, Ba^{2+}, Al^{3+}, SO_4^{2-} and SiO_2 in the brine. Traces of these impurities damage the membrane and/or electrodes and result in irrecoverable decreases in current efficiency and/or increased cell voltage. In the case of a closed brine loop with no purge, each impurity brought into or formed in the system must be removed to keep it below its specification level and to prevent accumulation.

The contaminants can be brought into the brine system by salt, by chemicals used in brine purification steps, by water for dissolving the salt, from materials of tanks, pipework, and cell components, or by the process itself [142]. The impurities in the salt depend upon the origin of the raw material. Rock salt, vacuum salt, sea salt, brine from well mining, or salt from waste incinerators serve as supplies of NaCl. The more varied the sources are, the more diverse the impurities.

Membrane and electrode damage effect cell performance, i.e., cause lower current efficiency, increased cell voltage, and, as a result, increased power consumption [143]. Some impurities affect the anode or cathode coating and cause an increase in overvoltage or simply deposit in the membrane, increasing its resistance and thus the cell voltage. The increase in voltage may in some cases be partially reversible when the impurity concentration drops to the recommended limits.

Table 14. Impurities and their effects

				Solubility in				Negative effect on performance Voltage increase					
Impurity	**Source**	**Max. limit (w/w)**	**Reagents**	brine	caustic	**Mechanism**	**Damage**	**An.**	**Cath.**	**Mem.**	**Ce**	**PQu**	**Methods of control**
Ca^{2+}	salt	Ca^{2+}+ Mg^{2+}	Ca^{2+}; OH^{-}	×	-	precipitation near the cathode	physical disruption of the membrane				++		precipitation with Na_2CO_3plus ion exchange
		< 20 ppb	Ca^{2+}; SO_3^{2-}	×	-	side of the membrane, formation					++		
			Ca^{2+}; SO_3^{2-}; AL^{3+}	×	-	of large crystals	(blister formation)				++		
			Ca^{2+}; I^{-}	-	-	very fine precipitation in the membrane	-						
Mg^{2+}	salt	Ca^{2+}+ Mg^{2+} < 20 ppb	Mg^{2+}; OH^{-}	×	-	fine precipitation near the anode side of the membrane	binding of ion-exchange sites in the membrane			++			precipitation with NaOH plus ion exchange

Table 14. (continued)

Impurity	Source	Max. limit (w/w)	Reagents	Solubility in		Mechanism	Damage	Negative effect on performance Voltage increase					Methods of control
				brine	caustic			An.	Cath.	Mem.	Ce	PQu	
Sr^{2+}	salt	< 4 ppm	Sr^{2+}; OH^-	×	×/-	precipitation on the cathode side of the membrane, formation of crystals	physical disruption of the membrane			+	++		coprecipitation with Na_2CO_3 plus ion exchange
Ba^{2+}	salt	< 0.5 ppm	Ba^{2+}; OH^-	×	×								purge;
			Ba^{2+}; I^-	×	-	very fine precipitation in the membrane	minor damage on the membrane, minor interaction with ion-exchange sites			+	+		precipitation with $NaHSO_3$ plus ion exchange
			Ba^{2+}; SO_4^{2-}			coating of the anode		+					

Table 14. (continued)

Impurity	Source	Max. limit (w/w)	Reagents	Solubility in		Mech-anism	Damage	Negative effect on performance Voltage increase			Ce	PQu	Methods of control
				brine	caustic			An.	Cath.	Mem.			
Al^{3+}	salt	< 0.1 ppm	Al^{3+}	×	×		-						precipitation as hydroxide at pH 7–9, ion exchange under acid conditions
			Al^{3+}; SiO_3^{2-}	×	-	formation of crystals (zeolites, sodalites, faujacites) near the cathode side of the membrane	disruption of the membrane				++		
			SiO^{3-}; Ca^{2+}	×	-	precipitation near cathode side of the membrane and crystalization	disruption of the membrane				++		Purge
			SiO^{3-}; Al^{3+}	×	-								

Table 14. (continued)

Impurity	Source	Max. limit (w/w)	Reagents	Solubility in		Mech-anism	Damage	Negative effect on performance Voltage increase					Methods of control
				brine	caustic			An.	Cath.	Mem.	Ce	PQu	
Hg^{2+}	parallel operation of an amalgam plant	< 0.2 ppm heavy metals	Hg^{2+}	×	×	deposition on the cathode	partially reversible, covering of active cathode coating		++				precipitation with Na_2S
Fe^{3+}	salt, pipework, tank material, anti-caking agent	< 0.1 ppm	Fe^{3+}	×	×	deposition on the cathode (in extreme cases: dendritic growth from cathode toward the anode)	covering of active coating (puncturing of the membrane)		++				precipitation with NaOH
Ni^{2+}	salt, pipework, tank material, cathode	< 0.2 ppm heavy metals	Ni^{2+}; OH^-	×	×	absorption of Ni in the membrane, deposition on the cathode			++	+			ion-exchange, purge

Table 14. (continued)

Impurity	Source	Max. limit (w/w)	Reagents	Solubility in		Mech-anism	Damage	Negative effect on performance Voltage increase					Methods of control
				brine	caustic			An.	Cath.	Mem.	Ce	PQu	
I^-	salt	< 0.2 ppm	I^-; Na^+	×	-	precipitation on the cathode side of the membrane, formation of crystals	physical disruption of the membrane				++		purge
			I^-; Ca^{2+}	-	-	very fine precipitation in the membrane	-						
			I^-; K^+	×	×		-						
F^-	salt	< 0.5 ppm		×			destruction of the anode coating	++					purge
SO_4^{2-}	salt, dechlorination with $NaHSO_3$	< 4–8 g/l Na_2SO_4	SO_4^{2-}; Na^+	×	-	precipitation near the cathode surface of the membrane	reduction of the OH^- ion rejection capability				++		purge, precipitation with $BaCO_3$ or $BaCl_2$ plus ion exchange
			SO_4^{2-}; Ba^+			coating of the anode		+					

Table 14. (continued)

Impurity	Source	Max. limit (w/w)	Reagents	Solubility in		Mechanism	Damage	Negative effect on performance Voltage increase					Methods of control
				brine	caustic			An.	Cath.	Mem.	Ce	PQu	
CO_3^{2-}	salt, precipitation with Na_2CO_3 or $BaCO_3$	< 0.4 g/l Na_2CO_3	CO_3^{2-}; H_2O			formation of CO_2						++	
ClO_3^-	process, side reactions	< 10 g/l $NaClO_3$				chlorination of ion-exchange resin	destruction of ion-exchaange resin						chlorate decomposition by acidification
TOC	salt	< 1 ppm				increased foaming, overplating		+	+	+			filtration
$K_4(Fe(CN)_6)$	anti-caking-agent for salt	< 0.1 ppm Fe^{3+}	Fe^{3+}	×		as Fe^{3+}			++				oxidation with active chlorine plus precipitation with NaOH
						evolution of N_2						++	

* An. = Anode; Cath. = Cathode; Mem. = Membrane; Ce = current efficiency; PQu = product quality.

Current efficiency declines are strictly related to the membrane. Impurities lower the current efficiency by reducing the membrane's ability to reject anions, specifically the ability to prevent hydroxyl ions from migrating from the cathode compartment through the membrane to the anode compartment [144]. This is usually a result of physical damage caused by precipitation and crystallization of impurities inside the membrane. Impurities precipitate because the environment in the membrane changes from an acidic salt solution (pH 2 – 4) to a caustic solution (pH 14 – 15) over the 100 – 300 μm thickness of the membrane.

It is important to consider not only the impurities themselves but also their interaction. The presence of one impurity may not be harmful, but its synergistic combination with others may be [144]. For example, silica itself is not harmful for membranes. Only in the presence of calcium and aluminum do precipitates form and damage the membrane irreversibly. The concentration of silica and/or the concentration of aluminum plus calcium can be adjusted to give the optimum operating conditions. For example, with an effective secondary brine purification, higher levels of silica can be tolerated. Similarly, if aluminum concentration is high, calcium or silica concentration must be reduced to maintain acceptable membrane performance.

To meet the strict requirements on brine purity outlined in Table 14 brine treatment is generally carried out in the following main steps in the brine loop: saturation, precipitation, clarification, filtration, polishing filtration, ion exchange, electrolysis, chlorate decomposition and dechlorination.

Calcium and magnesium are precipitated and separated from the saturated brine with the insoluble materials. A ca. 10 wt % sodium carbonate and barium carbonate (barium chloride) solution and 32 wt % caustic soda are used as precipitants.

$$Ca^{2+} + Na_2CO_3 \rightarrow CaCO_3 + 2\ Na^+$$

$$SO_4^{2-} + BaCO_3 \rightarrow BaSO_4 + CO_3^{2-}$$

$$Mg^{2+} + 2\ NaOH \rightarrow Mg(OH)_2 + 2\ Na^+$$

Alternatively:

$$SO_4^{2-} + BaCl_2 \rightarrow BaSO_4 + 2\ Cl^-$$

From the precipitation tank, the brine is fed into the clarifier, where a defined quantity of flocculant is added to promote the settling of the precipitated solids and gels. The brine is then pumped to a filtration system followed by an ion exchange purification.

Additionally, if the brine circulating system of an existing mercury cell plant also serves membrane cells, all mercury must be removed in a chemical treatment facility. The brine from the primary filtration is acidified with hydrochloric acid to pH 2.0 – 2.5 and sodium sulfide is added to precipitate mercury sulfide. Subsequently, the brine is filtered, alkalized to pH 9.5 – 11 by adding caustic soda, and finally fed to the secondary purification section.

The content of calcium and magnesium must not exceed 20 ppb. Such low contents can be achieved by using ion-exchange columns. The polished brine is pumped at approximately 70 °C to an ion-exchange system with two resins beds operating in series according to the lead/lag principle. When the leading ion exchanger is exhausted it is put to the regeneration and conditioning mode, while the lagging one takes over the lead position. After treatment in this secondary purification step, the purity limits are met and the brine is fed to the membrane cells. In another secondary purification system two columns operate in series while the third is in regeneration mode. When the first column is exhausted, the regenerated column is put in second position.

Prior to the resaturation, the byproduct chlorate and dissolved chlorine must be eliminated from the anolyte. Chlorate concentration is controlled by acidification of a partial stream of anolyte with an excess of hydrochloric acid [144]. Chlorine is removed under vacuum followed by addition of sodium bisulfite and hydroxide.

$$NaClO_3 + 6\ HCl \rightarrow NaCl + 3\ Cl_2 + 3\ H_2O$$

$$2\ Cl + 6\ NaHSO_3 + 6\ NaOH \rightarrow 4\ NaCl + 4\ H_2O + 2\ Na_2SO_4$$

All other impurities not precipitated, filtered out, or extracted by the ion exchangers can only be controlled by purging a partial stream of the anolyte to avoid accumulation. Resaturation then closes the loop.

7.2.2. Commercial Membranes

The ion-exchange membrane is the key component of the membrane cell. The energy consumption and the quality of the products depend on membrane performance. Requirements for the membrane are as follows:

1) Durability under the conditions of chlor-alkali electrolysis
2) High selectivity for sodium ion transport
3) Low electrical resistance
4) Sufficient mechanical strength for practical use
5) Low sensitivity to changing operating conditions

The importance of 1 – 3 is described in Section 7.1. High mechanical strength is necessary for installing the membrane and during service life, in which the membrane has to cope with deviations in temperature, concentration, and pressure.

As the performance of the membrane is the most important element in the economy of a membrane cell, many refinements have been made in membrane manufacturing. To reduce the current screening due to fabrics, membranes reinforced with dispersed

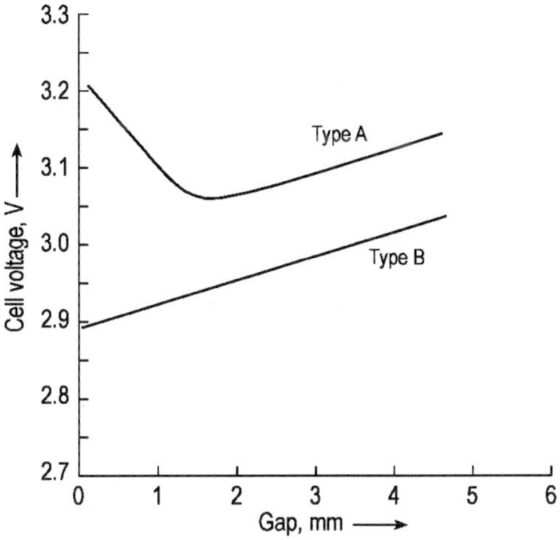

Figure 53. Effect of cathodic surface modification

microfibers and interwoven fabrics made of electrolyte-soluble fibers and PTFE have been developed [145], [146].

The improvement of hydrophilicity by covering the surface on the cathode side or on both sides with a nonconductive inorganic material brought about a significant reduction in the cell voltage. The surfaces of the membrane are covered with thin layers of a porous inorganic material. This material is an oxide, hydroxide, or carbide of the metals of groups 4, 5 and 6 or the iron triad (Fe, Co, Ni) [139], [147].

Figure 53 illustrates the effect of hydrophilic cathode surface modification. The surface-modified membrane (Type B) has a lower cell voltage than the conventional membrane (Type A). The voltage of the surface-modified membrane decreases linearly with decreasing gap size. With these advanced membranes, so-called zero-gap cells have been made possible, and the ohmic loss in electrolytes has been reduced to a minimum.

The active life of a membrane is determined by the economic balance between membrane cost and energy cost in use [149].

The performance of membranes depends on the operating conditions, especially on the caustic strength of the solution (Figure 52). Commercially available membranes, delivered by Asahi Chemical, Asahi Glass and Du Pont, are designated for use in a specific strength of caustic. For economic production the selection of the appropriate membrane is essential. Table 15 gives an overview of the most widely used membranes. Most membranes are operated in the narrow- or zero-gap configuration to minimize power consumption.

Table 15. Commercial membranes

		Available since	Tear strength, kg	Tensile strength, kg/cm	Ohmic drop (at 3 kA/m^2, $T = 90$ °C, $c = 32$ wt %), V	Caustic strength, wt %
Asahi Chem	F5201	1991	4	5.5	[a]	33 – 36
Aciplex	F4202	1993	4	5.5	[a]	30 – 34
	F4203	1997	4	5.5	[a]	30 – 34
	F890	1989	4.5	6	0.35	31.5 – 32.5
Asahi Glass	F892/old	1990	4.5	5	0.27	30 – 35
Flemion	F892/new	1994	4.5	5	0.28	31.5 – 33.5
	F893/new	1994	4.5	5	0.26	31.0 – 32.5
	N90209	1984	2.5	5.6	0.35	30 – 35
DuPont	N966	1988	5.5	7.3	[a]	30 – 35
Nafion	N981	1996	1.5	3.3	[a]	30 – 35

[a] Not published.

7.2.3. Power Consumption

For monitoring cell performance and comparing different electrolyzer designs, the electric power required to produce one tonne of NaOH 100 % is considered. This figure is determined by the voltage drop over one cell and the NaOH current efficiency.

$$\frac{\text{DC Energy}}{100\%\ \text{NaOH produced}} = \frac{U\,I\,t}{M} = \frac{U}{F\cdot\text{CE}}\left[\frac{\text{kWh}}{\text{t}}\right]$$

where
U is the cell voltage (V), F the Faraday constant for NaOH (1.4923 kg/kAh), and CE the NaOH current efficiency (%).

The specific power consumption is the main indicator for economic plant operation, and continuous efforts are made to lower the voltage and increase the current efficiency. At a thermodynamic minimum the decomposition voltage of about 2.2 V limits the theoretical minimum energy requirement to about 1480 kWh/t 100 % NaOH, as shown in Figure 54. At practical current densities of 3.0 – 5.5 kA/m^2 for present-day commercial cells and membranes, power consumption measured at the electrolyzer terminals is in the range of 1950 to 2180 kWh/t 100 % NaOH dependent on the selected current density (anolyte/catholyte temperature 90 °C, NaOH concentration 32 wt %, NaCl concentration in the anolyte 220 g/L). The power consumption rises with increasing operating time due to aging effects, such as decreasing current efficiency and increasing voltage. Investment costs rise when operating at low current densities, as more cells are needed to meet production. Hence electrolyzers are operated at low current densities in countries with high energy prices, and at high current densities in countries with low energy prices.

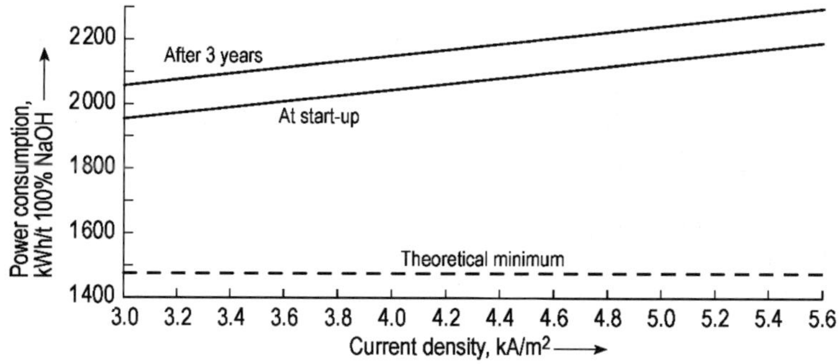

Figure 54. Specific power consumption

7.2.4. Product Quality

The caustic soda solution has a concentration of up to 32 ± 1 wt % NaOH. If a NaOH concentration of 50 wt % is required, evaporation can be used. The typical NaCl content is 20 ppm in a 32 wt % caustic solution.

The hydrogen has almost synthesis quality with a concentration of about 99.9 vol % H_2 (dry basis). The chlorine has an oxygen content of about 1.5 vol % (dry basis). Chlorine with an oxygen content below 0.6 vol % (dry basis) can be obtained by acidifying the brine with hydrochloric acid.

7.3. Membrane Cells

7.3.1. Monopolar and Bipolar Designs

A commercial membrane plant has multiple cell elements combined into a single unit, called the electrolyzer. The electrolyzers follow two basic designs: monopolar and bipolar [148].

In a bipolar arrangement the elements are connected in series with resultant low current and high voltage. The cathode of a cell is connected directly to the anode of the adjacent cell, as shown in Figure 55. The operation of a bipolar electrolyzer can be easily monitored by measurement of element voltages. If element upsets occur, a safety interlock system actuates the breakers (short-circuiting switches) and isolates the electrolyzer from the electric circuit. As the influx and efflux of electrolytes for the cells with different electric potential are gathered in common headers, problems of stray current may arise.

In the monopolar type all anodes and cathodes are connected in parallel, forming an electrolyzer with high current and low voltage (Figure 56). Due to the long current path, the voltage drop is high and can only be reduced by minimizing the size of cells or introducing internal copper conductors to lower the resistance. Because of this basic

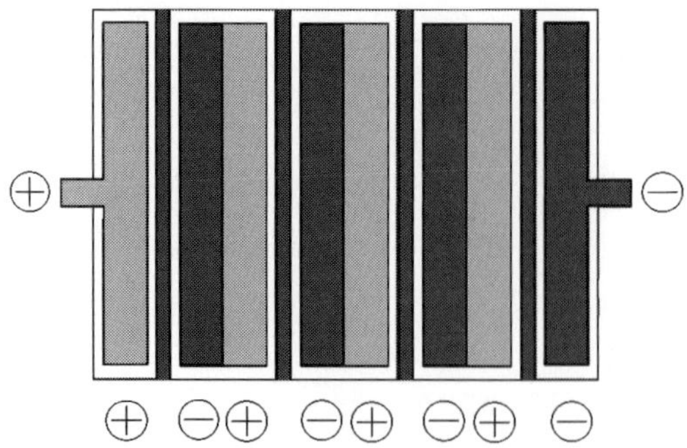

Figure 55. Bipolar electrolyzer

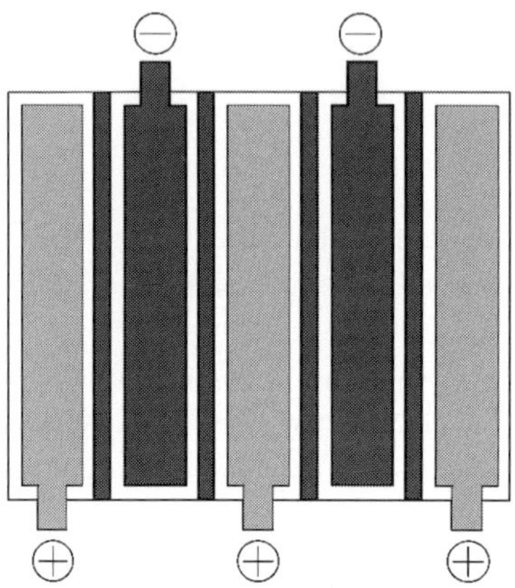

Figure 56. Monopolar electrolyzer

principle, ohmic losses in the monopolar cells are 80 – 100 kWh per tonne 100 % NaOH, which is much higher than in equivalent bipolar cells. Furthermore, the bipolar safety system is not applicable to the monopolar design, since the cell elements are arranged in parallel, which does not permit the monitoring of deviations in individual cell voltages.

Multiple electrolyzers are employed in a single d.c. circuit (Fig. 57). Usually bipolar electrolyzers are connected in parallel with low current and high voltage. Monopolar electrolyzers are often connected in series, resulting in a high current circuit and low voltage. Though both principles still appear on the market, investment and operating cost considerations, such as for the rectifier system, the cell room space required, for piping, valves, instrumentation, busbars and switches, significantly favor the bipolar design.

7.3.2. Commercial Electrolyzers

Generally, membranes are clamped vertically between the meshlike metal anodes and cathodes. The effective membrane area of a cell ranges from 0.2 to 5.0 m^2. Current density varies between 1.5 and 7 kA/m^2.

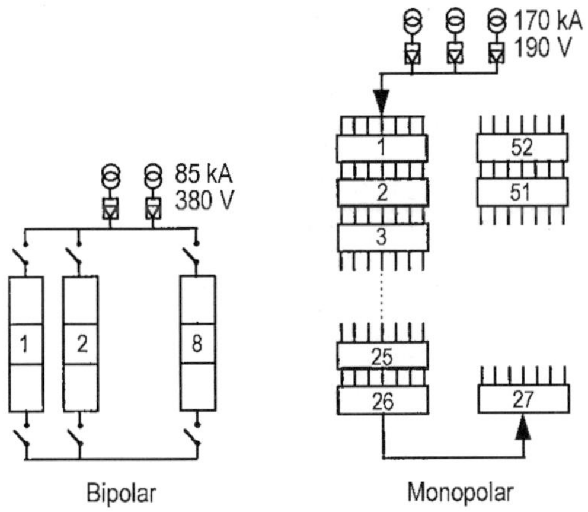

Figure 57. Electrolyzer architecture

The cells are filled with electrolytes, and gas-separating means are provided outside the cells. Many cells generally stacked like a filter press, constitute one electrolyzer with high production capacity.

The performance of a plant is determined by the electrolyzer, the cell voltages, and the current efficiency of the membrane. It is essential to design an electrolyzer with an homogeneous electrolyte concentration, temperature, and current density distribution across the whole area of the membrane.

The construction materials of the cell are selected to withstand the corrosive electrolytes. In most electrolyzers, titanium and nickel are used for the anode and cathode compartments of the cell. In older electrolyzers, stainless steel is used on the cathode side.

For economic and environmental reasons, mercury and diaphragm plants are increasingly being converted to membrane electrolyzers. The existing facilities, such as rectifiers, equipment for brine purification, and equipment for product treatment are utilized as much as possible.

Asahi Chemical ACILYZER-ML/NC Electrolyzer. The Asahi bipolar electrolyzer (Fig. 58) is of the filter-press type. The bipolar cell frames are suspended in a steel frame and compressed by a hydraulic device. Each cell frame consists of an anode and cathode compartment separated by a partition wall. The anode compartment is made of titanium, and the cathode compartment consists of special stainless steel and nickel. The anode and cathode structures are spot welded onto ribs in each compartment. Each compartment has an inlet nozzle for electrolytes at the bottom and an outlet nozzle for gas and electrolyte on top, connected to the gas/liquid separation chamber.

Two types of cell frames are available: frames with forced circulation of electrolytes by pumps, and frames with natural circulation in each compartment by means of a special arrangement of integrated ducts. The current is connected to the first and last element by flexible busbars.

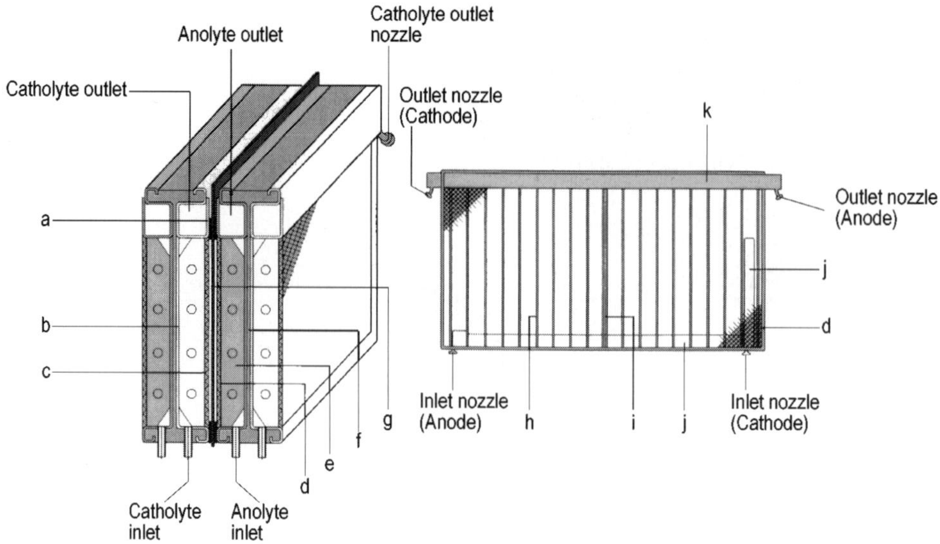

Figure 58. Asahi Chemical bipolar electrolyzer
Cell structure of ACILYZER ML32NC
a) Gasket; b) Nickel; c) Cathode; d) Anode; e) Titanium; f) Partition wall; g) Membrane; h) Rib; i) Reinforcing rib; j) Duct; k) Gas–liquid separation chamber

Asahi Glass AZEC-B1 Electrolyzer. The AZEC-B1 is a newly developed bipolar electrolyzer (Fig. 59). Metal bipolar cell frames are suspended in a steel structure similar to a filter press. A special hydraulic system presses the frames together. Caustic and brine are supplied by PTFE hoses to each frame, and discharge also takes place individually through PTFE hoses into metal headers. A special overflow method was developed for each frame to give smooth and stable electrolyte discharge.

Asahi Glass AZEC-F2 Electrolyzer. The AZEC-F2 (Fig. 60) is a monopolar metal-frame electrolyzer with a natural electrolyte circulation system. The monopolar metal frames are pressed together like a filter press between rigid end plates by long tie rods. The current is conducted to each frame from the intercell bus bar via flexible connectors. Gas and liquid leave the frame at the top and directly enter a gas separator box for anolyte/Cl_2 and catholyte/H_2. From these boxes, anolyte and catholyte are directly recycled through the feed headers into the respective compartments.

The Asahi Glass AZEC-M3 Electrolyzer. The AZEC-M3 is a monopolar rubber-frame electrolyzer with natural circulation of electrolytes. It was developed in 1981 and has been adapted to many clients in Japan and around the world. A simple electrode structure producing a minimal ohmic drop has been designed. The electrode sheets are clamped together with the membranes between rubber frames and gaskets. Manifolds for the inlet and outlet of electrolytes are formed in the frames and gaskets. The

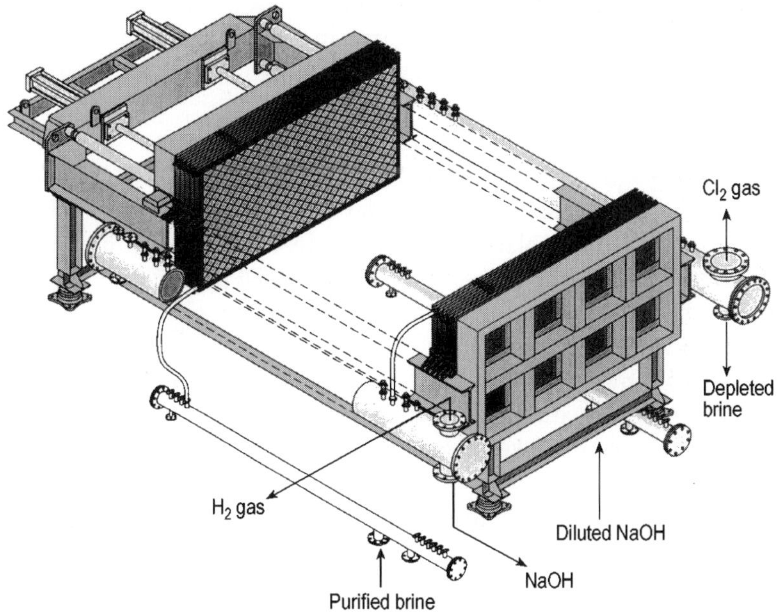

Figure 59. Asahi Glass AZEC-B1 electrolyzer

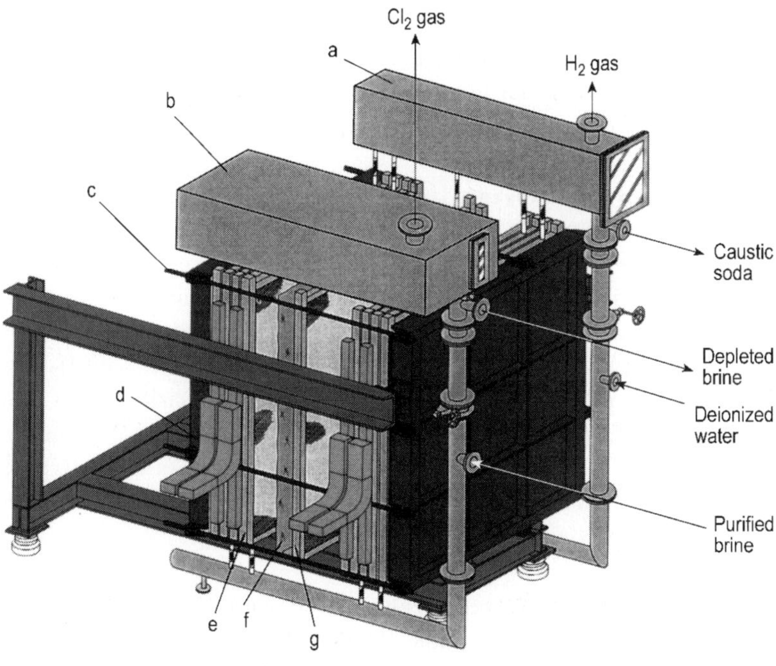

Figure 60. Asahi Glass AZEC-F2 electrolyzer
a) Cathode gas separator; b) Anode gas separator; c) Tie-rod; d) Flexible connector; e) Anode element; f) Membrane; g) Cathode element

discharge of electrolytes and gases takes place through inner ducts of the electrolyzer to gas–liquid separators. After separation from the gases, the electrolytes are recirculated through the electrolyzer as in the AZEC-F2 electrolyzer. The electrical connections between adjacent electrolyzers are kept short.

CEC BITAC 800 Electrolyzer. The Chemical Engineers Corporation (CEC) bipolar BITAC electrolyzer was jointly developed with Tosoh Corporation. The design follows the filter-press principle. Up to 80 bipolar electrode frames are clamped together by end plates and spring-loaded tie rods. The frames are made of special titanium alloy for the anode and nickel for the cathode. The electric current flows along the nickel pans, since the electrical conductivity of nickel is six times higher than that of titanium. Gas and electrolytes leave the cell compartment in overflow mode with little pressure fluctuation. Transparent PTFE tubes are attached at the electrolyte inlet and outlet nozzles of each element. Anolyte recirculation takes place through an external loop.

CEC CME DCM 400 Electrolyzer. The CME monopolar electrolyzer consists of large elements compressed in a filter-press arrangement. The electric current travels into each anode element through conductor rods and current distributors. This design achieves uniform current distribution over the large electrode area. The current distributors serve the additional role of a downcomer pipe, which creates a natural circulation within the cell, providing a uniformly distributed electrolyte concentration as well as good gas release. The anode frames are constructed from titanium and the cathode frames from a special stainless steel. The rods are cladded with titanium and stainless steel. Inlet and outlet tubes for liquids and gases and are made of transparent PTFE.

DeNora DN 350 Bipolar Electrolyzer. The DeNora DN 350 bipolar electrolyzers (Fig. 61) is of the filter-press type. The frames are very large (3.5 m^2) and are compressed by mechanical screw jacks to ensure proper tightness. The electrolyzer can be operated with a slight overpressure. Anolyte and catholyte are recirculated in the cell by baffle plates and special downcomers. Fresh brine and diluted caustic soda are fed to the bottom section of each compartment. Anolyte with Cl_2 and catholyte with H_2 leave the elements through insert pipes at the bottom of the cell. The electric current flows from the anode to the cathode surface through an array of welded electrical connectors of highly conductive metals (e.g., steel or copper). These connectors are interposed between the titanium anode wall and the nickel cathode wall, to which they are welded. The cathode walls are cold pressed to create a pattern of bulges.

DeNora D88/D175 Monopolar Electrolyzer. Monopolar electrolyzers with two different areas (0.88 and 1.75 m^2) are available. The DD-type monopolar electrolyzer (Fig. 62) is a self contained unit assembled on a separate base structure and fitted with an integrated gas–liquid system. The anodic and cathodic cell frames are compressed like a filter press with rubber gaskets and a tie-rod system. The current is conducted to

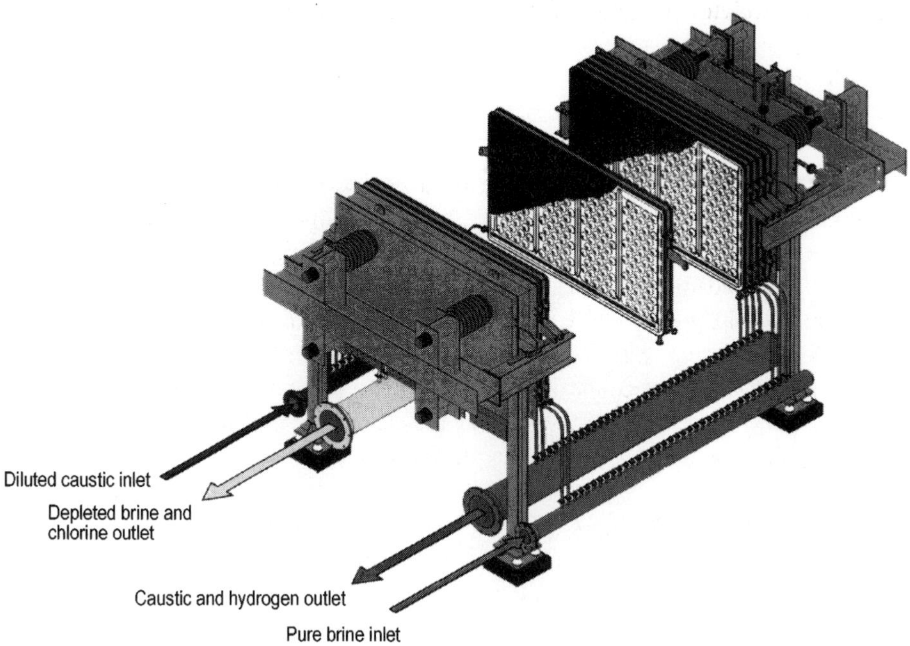

Figure 61. DeNora DN 350 bipolar electrolyzer

the frames by flexible copper busbars, and in the frame by cast-iron plate cladded with titanium and nickel. Anolyte/Cl_2 and catholyte/H_2 are led from the elements into the H_2 and Cl_2 disengagement boxes at the top of the electrolyzer. Recycling of anolyte and catholyte to the elements is driven by density differences. Feed brine and feed water are added to the cycle, whereas Cl_2/H_2 and the anolyte/catholyte surplus are withdrawn from the disengagement boxes.

Uhde BM 2.7 Electrolyzer. The bipolar Uhde electrolyzer (Fig. 63) is a single-element concept. Each element comprises anode and cathode half-shells, electrodes, a membrane, flanges, and the sealing system. This enables long-term storage of preassembled and fully tested elements. The electrodes are attached with continuous laser weld to the current transfer and support blades and hence to the half-shells. The anode is made of titanium and the cathode of nickel. The individual cell elements of an electrolyzer are suspended in a steel frame in which they are lightly pressed together for electrical contact. Large sealing forces are not required in the single-element concept, as each element is a separate, stand-alone electrolysis cell. The feed and discharge lines of the cell are located underneath the cells and connected to the catholyte and anolyte headers. The area above the electrolyzer is free of piping or bus bars, simplifying access and eliminating the risk of leakage and associated corrosion problems. The current is conducted from cell to cell by continuously laser-welded, explosion-bonded titanium –

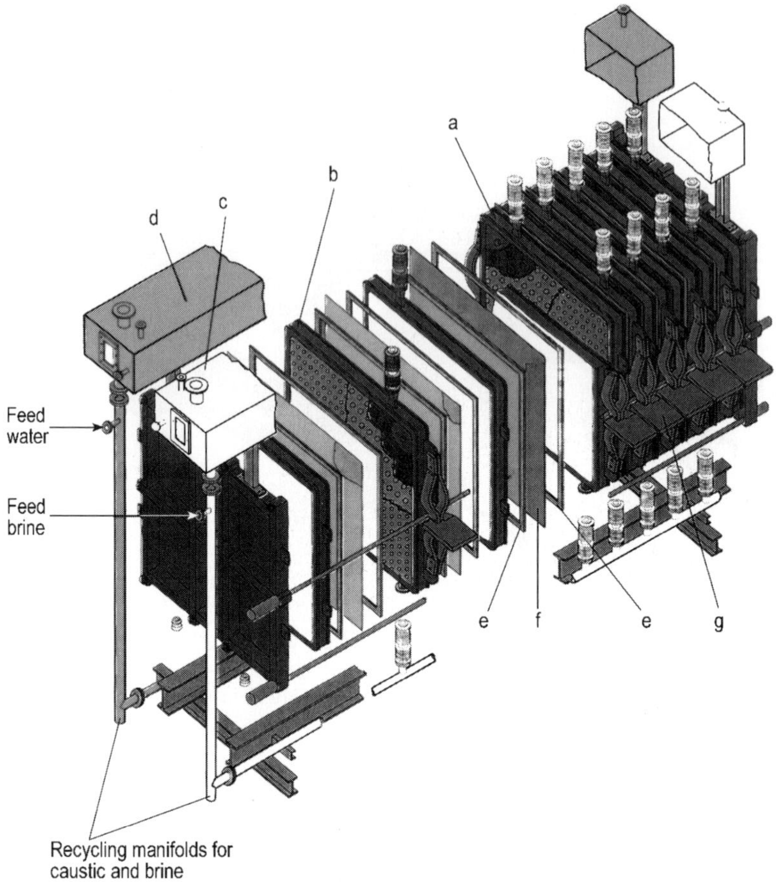

Figure 62. DeNora DD monopolar electrolyzer
a) Cathodic element; b) Anodic element; c) Chlorine disengaging box; d) Hydrogen disengaging box; e) Gasket; f) Membrane; g) Copper connection

nickel contact strips on the anode half-shell. The brine and caustic soda feeds enter the cell at the bottom, and the product streams are discharged downwards through internal overflow pipes. The internal baffle plate at the top of the anode half-shell prevents gas-phase blistering of the membrane. The chlorine gas is effectively removed from the membrane, preventing contact and improving the inherent safety of the electrolyzer. Natural circulation around a downcomer plate and a distribution pipe for brine and caustic achieve homogeneous temperature and concentration profiles within the element and assist in achieving uniform current distribution.

OxyTech ExL^{B} Bipolar Electrolyzer. The ExL^{B} bipolar electrolyzer (Fig. 64) is basically of the same design as the OxyTech ExL^{M} electrolyzer. Instead of the copper distributors with interface material, which provide the parallel (monopolar) arrange-

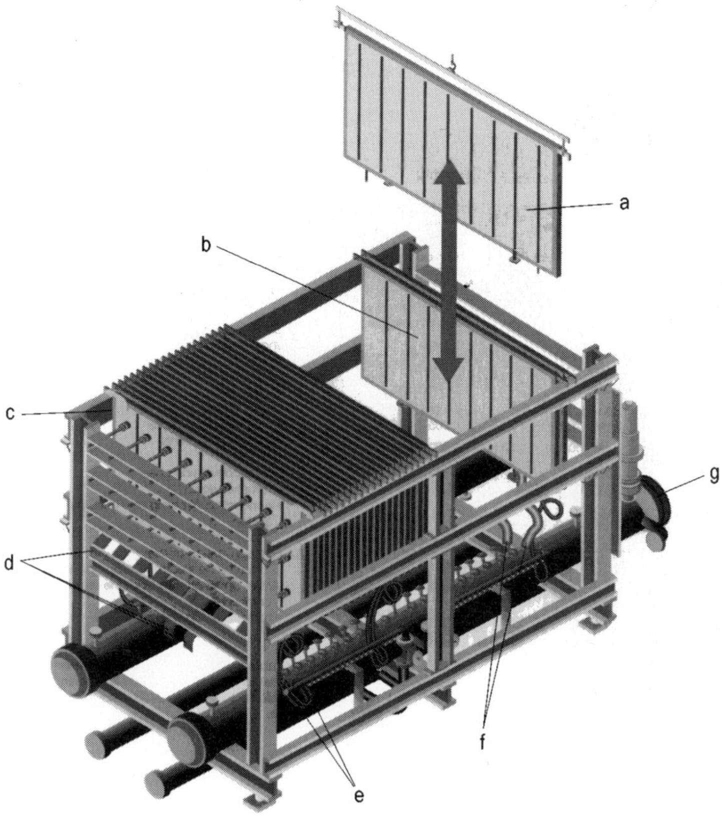

Figure 63. Uhde BM 2.7 electrolyzer
a) Single element; b) Contact strip; c) Cell rack; d) Busbars; e) Inlet hoses; f) Outlet hoses; g) Header

ment in the electric circuit, the elements are connected in series (bipolar), omitting the copper distributors and simply pressing the nickel cathode pan onto the nickel-plated back of the anode pan. The integral feed and discharge manifolds are designed to avoid current leakage.

OxyTech ExL^{M} Monopolar Electrolyzer. The ExL^{M} monopolar electrolyzer is an improved version of the MGC electrolyzer, which has been in service for more than 15 years. The elements are sealed with O-rings in a staggered gasket design. The cathode O-ring is located closer to the liquid than the anode O-ring. This protects the anode O-ring from the chlorination degradation, making it a long-life back up seal. The elements are pressed together by tie rods with copper distributor plates and conductive interface material to provide good current distribution. Electrolytes and gases are fed and discharged to and from the elements through the manifold passage attached to the cell elements. Increased internal electrolyte circulation is achieved by an improved electrode design.

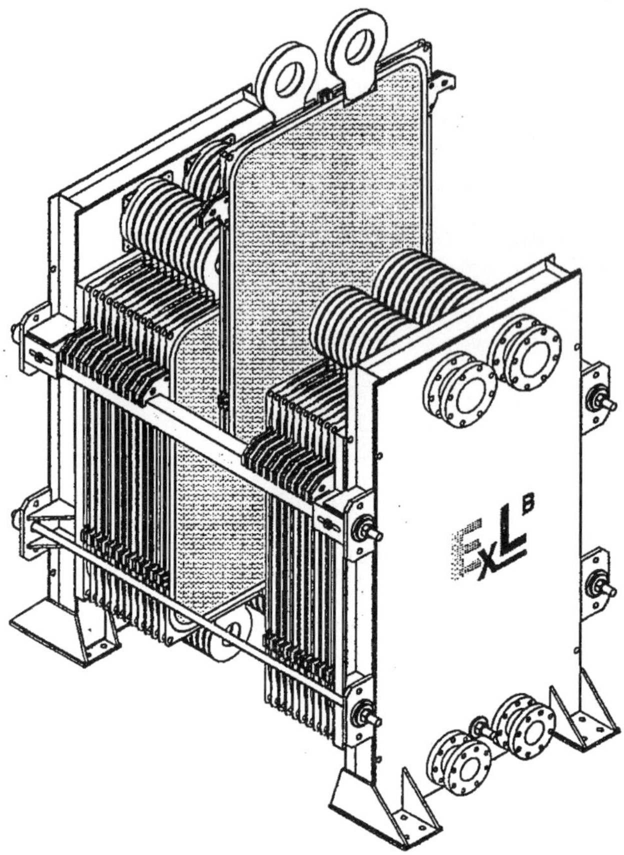

Figure 64. OxyTech ExL^B bipolar electrolyzer

OxyTech ExL^{DP} Dense Pak Unit Electrolyzer. The OxyTech ExL^{DP} dense pak electrolyzer comprises monopolar sections in one electrolyzer filter press compression set using standard monopolar cell components. Each monopolar cell section is separated by an insulating Inter Pak Spacer. Mostly three monopolar electrolyzer sections are included, with 2 to 10 elements per section. The dense pak can be configurated to match special rectifier/transformer configurations of existing plants, making it suitable for mercury and diaphragm cell conversion projects. For new plants, the advantage of the ExL^{DP} electrolyzer is reduced current, increased voltage circuits compared to an equivalent monopolar cell unit.

ICI FM21-SP Electrolyzer. The FM21-SP (Fig. 65) is a monopolar electrolyzer incorporating a simple pressed electrode structure. The anode assembly is composed of a 2 mm thick titanium panel between compression molded joints of a special cross-linked EPDM elastomer. The cathode assembly is composed of a 2 mm thick nickel panel between compression molded joints, also of EPDM.

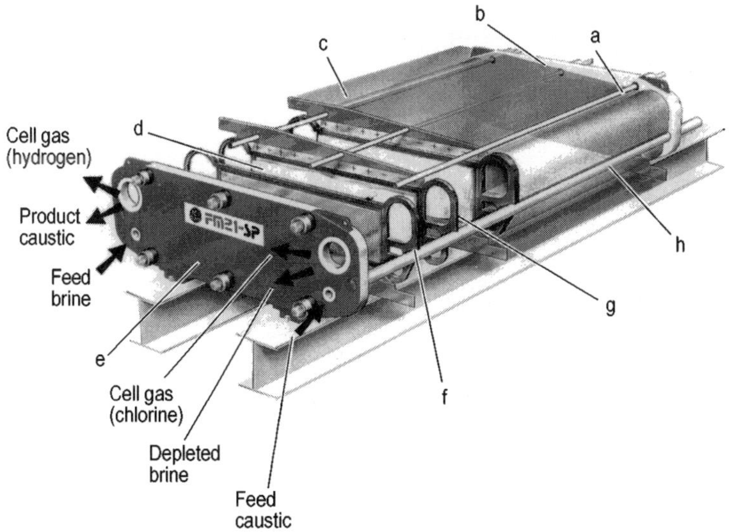

Figure 65. ICI FM21-SP electrolyzer
a) Tie-rod; b) Floating end plate; c) Copper electrical connections; d) Ion exchange membrane; e) Fixed end plate; f) Anode electrode assembly (titanium panel between compression molded gaskets; g) Cathode electrode assembly (nickel panel between compression molded gaskets); h) Support rail

The anodes and cathodes are assembled between 2 end plates until the number of electrodes required for the desired electrolyzer capacity is reached, up to 60 anodes in the FM21-SP and up to 90 anodes in the larger FM1500. A key feature of both designs is the elimination of any external piping to individual cell compartments by the use of a simple but effective internal header/manifold arrangement.

The electrolyzer has coated titanium anodes. The cathodes are pure nickel, also available with a coating to lower the hydrogen overpotential if necessary. Both electrodes are pressed from integral sheets of pure metal, and this makes recoating of the electrodes extremly simple and cost effective. Hence recoated structures can be sent to site prior to electrolyzer refurbishment from a pool of electrodes available to all customers.

Effective electrode area is $2 \times 0.21\ m^2$ per electrode, which gives a very compact electrolyzer. The individual electrodes are readily handled without the need for lifting apparatus, which allows the electrolyzer to be rebuilt and refurbished in the minimum of time.

7.3.3. Comparison of Electrolyzers

Operating parameters of bipolar electrolyzers are compared in Table 16, and those of monopolar electrolyzers, in Table 17.

Table 16. Bipolar electrolyzers

Company	ASAHI CHEM		ASAHI GLASS	CEC	DeNora	KRUPP UHDE	OXYTECH
Cell	**ML 32**	**ML 60**	**AZEC-B1**	**BITAC 800**	**DN 350**	**BM 2.7**	**ExL^B**
Effective membrane area m^2	2.72	5.05	2.88	3.276	3.5	2.72	1.5
Max. no. of elements	150	150	80	80	90	160	80
Current density, kA/m^2	up to 6.0	up to 6.0	1.5 – 6.0	1.5 – 6.0	up to 6.0	1.5 – 6.0	1.5 – 7.0
Max. capacity of electrolyzer t/d NaOH 100 %	45	90	48	54	65	90	29
d.c. power consumption kWh/t NaOH	2100	2100	2150	2150	2300	2130	2100
(at current density)	(4.0)	(4.0)	(5.0)	(5.0)	(6.0)	(5.0)	(5.0)

Table 17. Monopolar electrolyzers

Company	ASAHI GLASS		CEC	DeNora		ICI	OXYTECH
Cell	**AZEC F2**	**AZEC M 3**	**CME DCM 400**	**DD 88**	**DD 175**	**FM 21-SP**	**ExL^M**
Effective membrane area, m^2	1.71	0.20	3.03	0.88	1.75	0.21	1.5
Max. no. of elements	50	552	32	40	40	120	30
Current density, kA/m^2	1.5 – 4.5	1.5 – 4.0	1.5 – 4.0	3.5 – 4.0	3.5 – 4.0	1.5 – 4.0	1.5 – 6.0
Max. capacity of electrolyzer, t/d NaOH 100 %	13	15	13	4.8	9.6	7	9
d.c. power consumption kWh/t NaOH	2220	2046	2150	2300		2140	2150
(at current density)	(4.0)	(3.0)	(3.5)	(4.0)		(4.0)	(5.0)

7.3.4. Cell Room

Typical bipolar membrane cell rooms are shown in the following Figures 66 and 67.

The media are fed to and discharged from the electrolyzers by a header system arranged along the walls of the cell room on one side.

From the other side power is supplied either from separate transformer/rectifier units for each electrolyzer or from one unit for two or more (up to six) electrolyzers in parallel. The switches are arranged close to the rectifiers. They are actuated automatically and connected to the common interlock system for safety reasons.

In the middle space remains available for the electrolyzers and their individual feed and discharge piping. Only a light crane is required to handle single electrode frames or elements. Thus only a light structure for the entire cell house is used.

Figure 66. Bipolar cell room by Ashai Glass

Figure 67. Bipolar cell room by Krupp Uhde

8. Electrodes

8.1. Anodes

The initial anodes used for the electrolytic generation of chlorine were made of platinum or magnetite. However, as the plant grew in the size, the cost of platinum and limitations of the current density for magnetite led to the wide-scale introduction of graphite anodes, which were used exclusively up to 1970. The graphite of choice was low in ash and vanadium and composed of various types of particulate coke and pitch binder. Following extrusion, baking at ca. 1000 °C, and graphitization at 2600 – 2800 °C, the final shape of the electrode was achieved by machining. The shape of the horizontally suspended anodes with an initial thickness of 7 – 12 cm for the amalgam process was similar to that of modern titanium anodes due to the retrofitting of existing cells. The anodes had vertical slits and holes to allow the removal of the gaseous chlorine. Due to the cogeneration of oxygen and the resulting formation of CO and CO_2, electrode wear was high, in the range of 1.8 – 2.0 kg graphite per tonne of chlorine from NaCl and 3 – 4 kg per tonne from KCl. Even with a daily adjustment of the anodes to compensate for the changes in dimension a k value of only 0.12 to 0.14 $Vm^2\ kA^{-1}$ was achievable.

The initial attemps to replace the graphite anodes with activated titanium anodes began as early as 1957 with platinized titanium and Pt/lr-coated anodes. However because of the short lifetimes of the anodes, they were not economic. The use of mixed metal oxides was first patented by Beer in 1965 and 1967 [150]. The initial patent described a coated metal electrode in which the active material was a mixed metal oxide coating containing one or more of the platinum metal group oxides. The second patent described coatings in which mixed metal oxide crystals contained a non-platinum metal oxide in addition to the platinum metal oxide (including Ti, Ta, and Zr oxides).

Further improvements in the coating and the anode structures followed rapidly along with the commercialization of anodes by De Nora [151] under the trade name Dimensionally Stable Anode (DSA). Because of the dimensional stability and the lifetime of the coating and the ability to increase the current densities, rapid introduction of the activated titanium anodes was possible. At present only a few plants still use graphite anodes, largely due to the initial investment costs for titanium anodes.

8.1.1. General Properties of the Anodes

Coating Properties and Preparation. Comprehensive reviews on preparation and properties are given in [152], [153].

Chemical Composition. Because of its price and performance, Ru is the basic component in all commercial coatings at present, along with an oxide of a non-platinum metal (e.g., Ti, Sn, or Zr). In most cases a second platinum metal oxide is added to increase the performance of the anode coating. There is an optimum ratio of platinum metal oxides to non-platinum metal oxides in terms of overpotential, wear rate, and costs. The optimum depends on the operating conditions and the method of preparation of the coating and normally lies in the range of 20 : 80 to 55 : 45 by weight. Some of these coatings may contain glassy fibers [154] and some contain pre-oxidized material such as $Li_{0.5}Pt_3O_4$ [155].

Preparation. The solvent used for the preparation of the precursors solutions are chosen on the basis of the desired electrochemical properties and the method of application, which is mainly determined by the anode structure. Most coating solutions are prepared by dissolving salts or organometallic complexes in aqueous, organic, or mixed solvents. The coating can be applied by spraying, brushing, dipping, or other techniques. Following evaporation of the bulk of the solvent, the anode is heated to 350 – 600 °C to form the oxidic coating then cooled prior to the next coating cycle. This is repeated until the desired coating thickness is applied. Optional post-thermal treatment can also be carried out.

The optimal performance of the coating depends on the above parameters and on the coating thickness per coating cycle, which must be optimized for each coating and surface pretreatment step.

Crystallographic Composition, Morphology, and Real Surface Area. A rutile phase is the electrochemically active phase of the coating, and although it is thermodynamically unstable, it remains even after many years of operation. The stable phase — anatase TiO_2 in the case of $TiO_2 – RuO_2$ coatings — is electrochemically inactive [156]. The degree of crystallinity and the composition are related to the processing parameters [157] and the various degrees of mixed crystals exhibit different stabilities. The real surface area of the coating is a function of both the titanium pretreatment and the coating composition. The surface of the chlorine-generating coating is often described as "cracked mud" due to its resemblance to a dry river bed. The BET surface area of the coatings or that determined electrochemically vary ca. 400 to 1000 times the geometric surface area [158].

Overpotential and Current – Voltage Relationship. The observed overpotential for chlorine evolution at 2 – 10 kA/m^2 is in the range of 80 – 110 mV [159] – [162], about 70 – 100 mV of which is due to diffusion overpotential effects [162]. The overpotential for the generation of oxygen under similar pH and temperature conditions lies is ca. 300 mV more anodic than that of chlorine generation. Other than oxygen evolution, the only other side reaction is formation of chlorate.

Coating Wear and Coating Lifetime. The coating lifetime is strongly dependent not only on the type of cell — membrane, diaphragm, or mercury — but also on a range of process parameters, including brine quality, current density, and membrane or

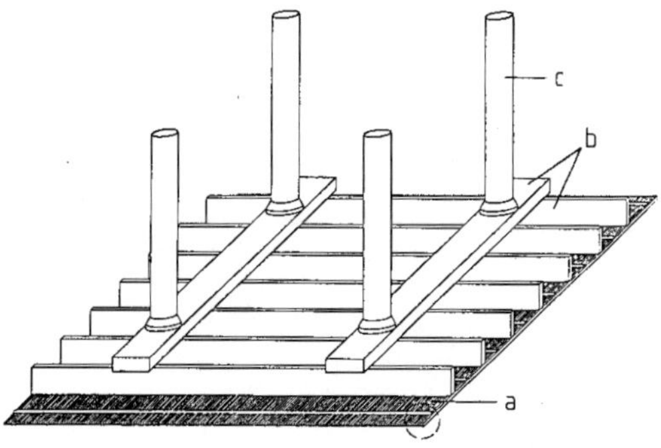

Figure 68. Four-stem anode for amalgam cells a) Active surface; b) Current distributor; c) Riser tube to protect the copper bar inside

diaphragm quality. The upper limit of the wear rate would seem to be in the region of 500 t Cl_2/m^2 anodic area for a standard commercial loading.

The wear rate mechanism is discussed in detail in [163], [164]. The effects of various impurities and materials in the brine can be divided into three types.

1) Compounds or ions which attack the substrate, e.g., fluoride or organic acids such as formic or oxalic acid.
2) Materials which built up blocking layers on the surface of the anode, e.g., hydraulic oil or polymer films resulting from delamination of membranes. The irreversible poisoning of coating is caused by ultrathin aluminum silicate layers.
3) Electrochemically active film-forming materials such as MnO_2, which may lead to an increased oxygen content in the chlorine.

Other examples, such as the insensitivity of the performance of diaphragm anodes to almost complete surface coverage by iron oxides illustrate the robustness of commercial coatings.

8.1.2. Anodes for Mercury Cells

Structure. The classical structure of anodes for this process still reflects the retrofitting concept used during the 1970s and the high current operations at ca. 10 kA/m^2. A typical mercury cell anode consists of a number of copper shafts, protected by either a permanently welded or removable titanium outer sleave, from which the current is distributed to the active surface over distributor bars (Fig. 68).

The quick release of gas and the supply of fresh brine to the active surface are the major requirements of an mercury cell anode, and a wide range of designs have been built. The most common types are shown in Figure 69. The differences are more evident at current densities > 7 kA/m^2. The use of baffles on the back of the active surface to enhance the gas lift and aid the supply of brine to the active surface is also common [165].

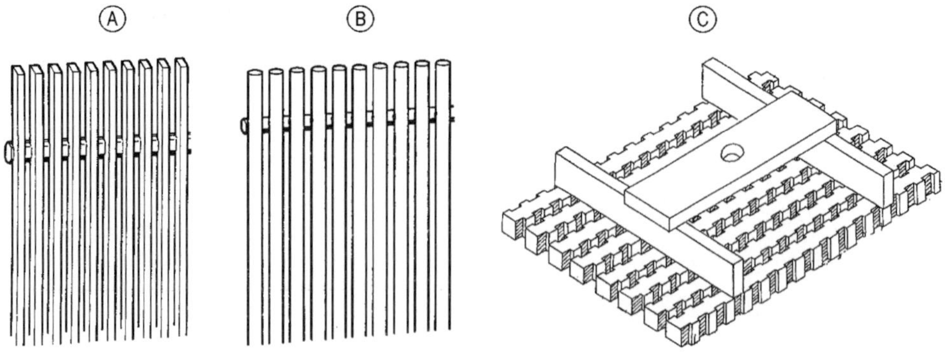

Figure 69. Anode designs for quick gas release
A) Flat profile (channel blades); B) Rod type (3-, 4-, 5-mm diameter); C) 3D side profile anode

Flatness is critical for the optimal performance of the anodes in the cells. A typical value < 0.5 mm is achieved, mostly by manual straightening after manufacturing or recoating.

Coating Life. The coating life is determined by a wide range of practical aspects and normally not directly related to the electrochemical wear rate of the coating. These include:

- Mechanical damage to the anode caused by short circuiting [166], [167]
- The need to maintain a recoating schedule, due to production demands and the labor intensive refitting of a cell.
- Synchronization of recoating with the exchange of other consumable parts of the cells such as covers and gaskets.

The tendency in recent years has been to increase the lifetime from about 180 t Cl_2/m^2 to 300 – 400 t Cl_2/m^2. This has been achieved by the introduction of better control systems in the cells and the development of intermediate layers of plasma-sprayed conductive TiO_{2-x} between the active coating and the titanium substrate [160].

8.1.3. Anodes for Diaphragm Cells

The predominate determinants in the design of diaphragm anodes are:

- The relatively low current density of ca. 2.0 kA/m^2
- The minimization of the anode – diaphragm gap
- The need to remove and replace the anode array from the cathode, hence the use of retractible anodes

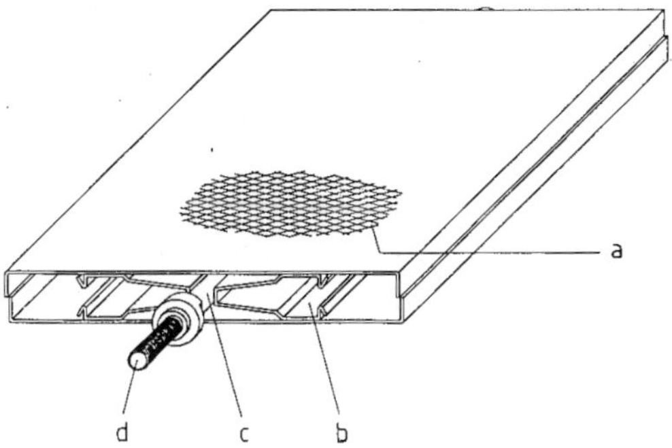

Figure 70. Anode for monopolar diaphragm cells
a) Activated (coated) expanded metal; b) Expanding spring; c) Titanium-clad copper bar; d) Copper thread to fix the anode to the cell base

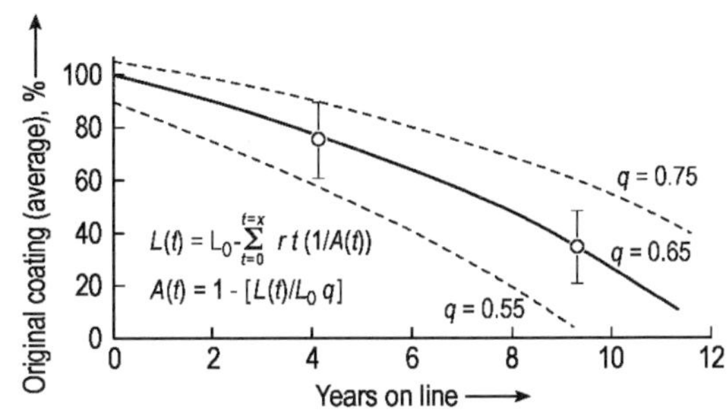

Figure 71. Empirical fit of observed nonlinear wear rate of coating thickness versus years on line. *L*(*t*) is the loading at time *t*, *L* the initial loading, *r* the wear rate, t time, *A*(*t*) the active surface area at time *t*, *q* an empirical factor related to the current density sensitivity of the wear rate *r*

- The limiting height of the cathode, integrated with the diaphragm manufacturing technology

In the recent past, the conventional anodes shown in Figure 70 [168] have been further developed to optimize the energy consumption of the cells by replacing the simple flat expanded metal with complex structures [170]. At present very few plants are still operating without expandable anodes.

Another type of diaphragm anode is used in the bipolar Ganor cells [169].

Coating Life and Mechanism of Deactivation. The coating lifetime of DSA coatings exceeds 12 years, and production of chlorine exceeds 240 t Cl_2/m^2. The wear is caused by the relatively high oxygen content in a diaphragm cell [171] of ca 1 – 2 %. The wear rate is nonlinear (Fig. 71). This nonlinearity is critical for determining the correct time to begin recoating so as to avoid unplanned stoppages.

8.1.4. Anodes for Membrane Cells

Structure. The variety of designs of membrane cells has led to a range of anodes active area structures; the common principles are the need to support the membrane and gas release to the back of the anode surface. Therefore, thin flattened expanded, perforated metals or louver type structures with and without perforations are used [172].

Coating Life. At present the second-generation coatings for membrane cells are showing lifetimes comparable to those of the diaphragm process. The actual lifetime of the anodes is dependent on the extent of damage by caustic flow through holes in the membrane or by contamination with the poisons.

Oxygen Content [152]. The oxygen content has also been improved by the optimization of internal circulation of the brine within the cells.

8.2. Activated Cathode Coatings

Since 1910 diaphragm brine electrolyzers have used carbon steel cathodes and continue to use carbon steel to this day. When the first ion-exchange membrane electrolyzers were introduced in the late 1970s, the cathodes were also carbon steel. By the early 1980s the design had evolved to stainless steel and nickel cathodes, and finally in the 1990s to exclusively nickel cathodes. Depending on current density, the hydrogen overpotential of carbon steel cathodes is about 300 mV. Active cathode coatings can lower the overpotential by 200 – 280 mV, thus providing significant energy savings. Active coatings have often been described in the literature and used in water electrolysis for over 40 years. With the development and evolution of the ion-exchange membrane technology, active cathode coatings are coming into general use.

The patent literature covers many different types of coatings, and new ones are being published regularly. The two basic approaches to activation are high-surface area coatings and catalytic coatings. Both bare nickel and carbon steel show lower hydrogen overpotential once in operation and their surfaces roughen. In fact by grit blasting bare nickel cathodes and roughening the surface, the long-term overpotential can be reduced by 30 – 40 mV. More common are porous nickel-type coatings that offer high surface area and good chemical resistance. These coatings consist of two or more components. At least one of the components is leached out in caustic to leave the porous high surface area nickel [173]. These coatings are typicall nickel-zinc [174], nickel – aluminium – Raney nickel [175], nickel – aluminium [176], or nickel – sulfur [177]. A variety of additives are recommended for strength, life, and resistance to poisoning by impurities. Rough coatings of nickel – nickel oxide mixtures [178] and nickel with embedded activating elements such as ruthenium [179] are also used. Sintered nickel coatings are described in patents [180] as well as being available from Huntington Alloys. Nickel coatings containing platinum group metals, primarily platinum and/or ruthenium, have been sold by Dow [181], Johnson Matthey, and ICI [182].

The coatings used for diaphragm and membrane electrolyzers differ because of the different substrates (carbon steel and nickel, respectively) and the different operating conditions. The weak 11 % caustic in diaphragm cell liquor is less corrosive than the strong 33 % caustic of a membrane electrolyzer. The less expensive and more fragile coatings like nickel–zinc can be used in diaphragm electrolyzers. Membrane electrolyzer suppliers favor the platinum group metal coatings.

The shape of the cathode structure is an important factor affecting the choice of cathode coating. The complex cathodes of diaphragm electrolyzers lend themselves to liquid systems (e.g., electroplating or electroless baths) that can coat the entire structure by immersion [183]. Membrane electrolyzers, which are primarily of a filter-press design, have flat cathodes that are easy to coat by spraying or painting. The lower operating current density of diaphragm electrolyzers means more cathode area per unit of production; this requires a less expensive coating. Most diaphragm electrolyzers use heat-cured polmer–asbestos separators (diaphragms) that are vacuum deposited after the cathode coating is applied. This curing operation can destroy the activity of certain coatings.

All cathode coatings are susceptible to poisoning by impurities that make their way into the catholyte with the deionized water or are components of the piping, electrolyzer etc. These impurities tend to blind the activity of the coatings over a period of time that depends on their concentration. Porous nickel coatings in diaphragm electrolyzers are less susceptible to blinding by impurities because spalling of the brittle coating makes the coatings self-cleaning. The platinum group metal coatings are subject to damage from reverse currents during electrolzer outages. Precautions are needed to protect the coatings with reducing agents [184] or by cathodic protection [185].

Active cathode coating have become the standard throughout the chlor-alkali industry for new construction with the ion-exchange membrane electrolyzer technology. In most of the older diaphragm electrolyzer plants, problems with application of the cathode coatings and generally lower power costs have obviated the use of active cathode coatings. While there are more recent developments applicable to the diaphragm technology in the way of active cathode coatings, many of these developments remain the proprietary information of the technology and coating suppliers.

9. Comparison of the Processes

The advantages and disadvantages of the three chlor-alkali processes are summarized in Table 18. The three chlor-alkali processes can be compared in respect to the quality of the chlorine and caustic produced, and the equipment and operating costs.

Today the membrane process is the state of the art for producing chlorine and sodium hydroxide or potassium hydroxide. All new plants are using this technology. The production capacity of chlor-alkali plants using the membrane process reached about 21% of total world production capacity in 1995 and is predicted to increase to about 28 % by 2001 (Table 19) [133]. The diaphragm cell capacity remains constant and there is a decline in mercury cell capacity.

The conditions for a conversion from the mercury and the diaphragm process to the membrane process are discussed below.

9.1. Product Quality

Table 20 shows typical composition values for the chlorine and caustic produced by the diaphragm, mercury, and membrane processes. Chlorine produced by the mercury process can be used directly for most uses. Chlorine produced by the diaphragm or

Table 18. Advantages and disadvantages of the three chlor-alkali processes

Process	Advantages	Disadvantages
Diaphragm process	use of well brine, low electrical energy consumption	use of asbestos, high steam consumption for caustic concentration in expensive multistage evaporators, low purity caustic, low chlorine quality, cell sensitivity to pressure variations
Mercury process	50 % caustic direct from cell, high purity chlorine and hydrogen, simple brine purification	use of mercury, use of solid salt, expensive cell operation, costly environmental protection, large floor space
Membrane process	low total energy consumption, low capital investment, inexpensive cell operation, high-purity caustic, insensitivity to cell load variations and shutdowns, further improvements expected	use of solid salt, high purity brine, high oxygen content in chlorine, high cost of membranes

Table 19. World chlorine market 1995 and 2001 (in %)

	1995	2001
Diaphragm	52	49
Mercury cell	22	18
Membrane	21	28
Others	5	5
Market, 10^6 t/a	41	43.3

Table 20. Product qualities: typical compositions of chlorine, caustic, and hydrogen

Product and contents	Process			
	Diaphragm		Mercury	Membrane
	Unpurified	Purified		
Chlorine gas (from cells), vol %				
Cl_2	96.5 – 98		98 – 99	97 – 99.5
O_2	0.5 – 2.0		0.1 – 0.3	0.5 – 2.0
CO_2	0.1 – 0.3		0.2 – 0.5	
H_2	0.1 – 0.5		0.1 – 0.5	0.03 – 0.3
N_2	1.0 – 3.0		0.2 – 0.5	
NaOH solution, wt %				
NaOH	50.0	50.0	50.0	50.0
NaCl	1.0	0.025	0.005	0.005
Na_2CO_3	0.1	0.1	0.05	0.04
Na_2SO_4	0.01	0.01	0.0005	0.0001
$NaClO_3$	0.1	0.001	0.0005	0.001
SiO_2	0.02	0.02	< 0.001	0.002
CaO	0.001	0.001	0.001	0.0001
MgO	0.0015	0.0015	0.0002	0.0001
Al_2O_3	0.0005	0.0005	0.0005	0.0001
Fe	0.0007	0.0007	0.0005	0.0004
Ni				
Cu	0.0002	0.0002	0.00001	0.0001
Mn				
Hg	none*	none*	0.00001	none*
NH_3		0.001		
Hydrogen gas, vol %				
H_2	> 99.9		> 99.9**	> 99.9

* $< 10^{-6}$ %.

** Hydrogen gas from the mercury process contains mercury: 1 $\mu g/m^3$ – 10 mg/m^3, depending on the purification process. The hydrogen gas from the other two processes is free of mercury.

membrane process contains up to 2 % O_2, depending on the pH of the anolyte. This oxygen can be removed by condensation and evaporation of the chlorine.

The sodium hydroxide solution from the mercury process is the purest of the three; the amounts of NaCl and $NaClO_3$ are especially low. However, the quality of caustic from the membrane process is almost as good. A main drawback of the diaphragm process is the high concentration of NaCl and $NaClO_3$ in the caustic solution. This sodium hydroxide solution cannot be used for some processes. A chloride-free grade, commonly referred to as rayon-grade caustic, is required for 20 – 30 % of the demand in industrialized countries. Even the use of purification processes (see p. 73) does not reduce the NaCl content below 0.03 wt %. In addition to the NaCl and $NaClO_3$, the levels of Si, Ca, Mg, and sulfate impurities are higher than for the mercury and membrane processes.

9.2. Economics

The wide variation in the main cost factor, that for electrical energy, which varies from region to region by a factor of up to three, makes a direct comparison of production costs problematic. Further, the cost of electrical energy is increasing in different regions at drastically different rates, depending on the basic source of energy and customs. Rapidly changing foreign exchange rates also make international comparisons difficult.

A detailed discussion of the capital investment and operating costs for the three processes for a 200 000 t/a-plant in 1991 is given in [186]. A comparison of the investment costs does not make sense today for the mercury process, because no mercury cell plant and only a few diaphragm cell plants were built since then. All new plants are using the membrane process.

9.2.1. Equipment

The expenses for the rectifier, chlorine and hydrogen systems, HCl system, caustic storage, utilities, and engineering and construction overheads are approximately the same for the three processes.

Cells. The complex mercury cells are considerably more expensive than the simpler diaphragm and membrane cells. There is no development in mercury cell technology. Improvements are being made in diaphragm cells (higher current densities, longer service times), but the relative advantage of the membrane cells is rising fast with considerable increase in current density and improved membrane performance: Fewer cells are needed for a given production capacity.

Brine System. The brine system for the diaphragm process is the simplest of the three — there is neither dechlorination nor sulfate precipitation, except in some very specific cases — and makes up only 3 – 4 % of the capital investment. The brine system is the most complex for the membrane process, for fine purification by ion exchange is necessary. However, the two- or three-fold greater depletion of the brine in the membrane process allows the brine system to be smaller than that for the mercury process. Therefore, the cost of the brine system for either process is approximately the same, 4 – 7 % of the total.

Caustic Concentration. The elaborate multistage evaporators required for the concentration of the diaphragm-cell caustic and the separation of NaCl and Na_2SO_4 must be nickel plated because of the corrosiveness of the cell liquor containing NaCl and $NaClO_3$. These evaporators cost 20 – 35 % of the total. The evaporators for the membrane process may be constructed of stainless steel and are much smaller because the essentially salt-free cell liquor is more concentrated, costing 3 – 4 % of the total. The mercury process produces 50 % caustic directly, evaporation is not required.

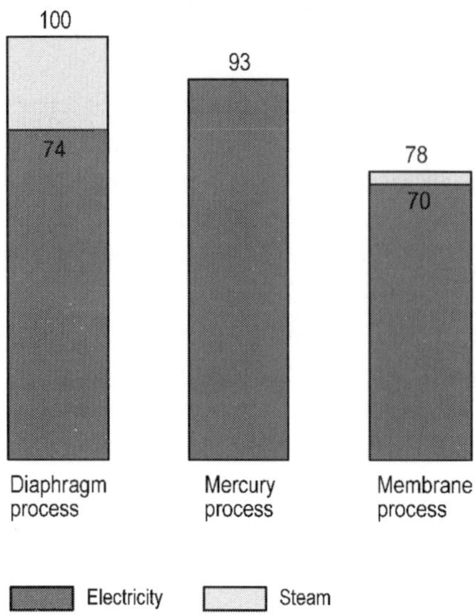

Figure 72. Relative consumption of energy (electricity and steam) in the three chlor-alkali processes in producing 50 wt % NaOH

Facilities for Handling Salt. The mercury and membrane plants require storage and handling facilities for solid salt. If a diaphragm plant uses well brine, only small facility is needed for the recycling of the salt from the caustic evaporation.

Mercury. In addition to the capital cost of mercury itself, there is the expense of the equipment to prevent emission of mercury into the environment and to remove mercury from the products (see 5.3.5). This equipment costs 10 – 15 % of the total capital investment.

The investment cost of a new (green-fields) chlor-alkali project in the USA is estimated to be between 250 000 and $ 300 000 per tonne per day chlorine capacity in 1998 [187].

9.2.2. Operating Costs

The fixed costs for operators and other personnel, taxes, insurance, repairs, and maintenance are about the same for all three processes. The 20 % lower depreciation of the membrane process is offset by the additional expense for purchase and replacement of the membranes and for the more elaborate brine purification.

Of the variable costs, the expense for salt, precipitants, and anode reactivation are roughly the same. The difference among the three processes shows up in the consumption of energy, as electricity and steam. If 1 t of steam is taken to be equivalent to 400 kW h of electrical energy, then the comparison in Table 21 can be made. The differing total energy consumptions are illustrated in Figure 72.

The price of electrical energy varies widely from region to region. The relatively broad range of possible current densities combined with the steep increase in the cell

Table 21. Energy consumed to produce 1 t of chlorine plus 1.13 t of caustic soda (50 %) in the three chlor-alkali processes

Energy	Process		
	Diaphragm	Mercury	Membrane
Electricity for electrolysis, kW h	2300 – 2900	3100 – 3400	2200 – 2600
Steam equivalent, kW h	800 – 1000	0	200 – 400
Total, kW h	3100 – 3900	3100 – 3400	2400 – 3000
Relative energy costs	100 %	93 %	78 %

voltage with current density for the diaphragm and membrane cells allows optimization of the current density with respect to the local energy price. That is, if electrical energy is relatively expensive, a greater number of cells, and thus a greater capital investment, can be tolerated to reduce the specific energy consumption and thus minimize total unit production cost [188].

9.2.3. Summary

In spite of the advantages of the membrane technology, about 75 % of all chlorine is produced in mercury and diaphragm cells, operating in ca. 500 plants around the world. Diaphragm technology prevails in the United States (70 %), Russia, and China, and mercury technology in Western Europe (64 %). Continued production from these plants is economical under special circumstances.

For mercury cell users, the question of today is whether the old, depreciated plant is competitive with new membrane cell plants. The alternatives are:

- Further production in the mercury cell plant
- Conversion to the membrane process
- Phasing out the old plant

Candidates for further production are plants of medium to large size, with low electricity costs, with very high quality products, with high emissions standards, with high maintenance standards (low repair costs), or which produce speciality products which cannot be obtained in membrane cells (e.g., alkoxides or dithionites).

All producers that do not fulfill one or more of these conditions are candidates for conversion. The more the existing infrastructure can be used, the greater the benefits resulting from conversion. The investment for the conversion of a middle-sized plant (100 000 t/a) is between \$ 550 and \$ 800 per tonne of chlorine capacity per year [189]. This investment includes the cost for the membrane cells, secondary brine purification and additional changes of the infrastructure. The costs are specific for each existing plant and depend on:

- Chlorine quality (e.g., the oxygen content)
- The use of existing buildings. The materials of the existing brine treatment area
- The possibility of using the dilute caustic (32 – 36 %) within the plant without concentration
- Use of the existing electrical equipment, rectifiers, busbars
- Possible capacity enlargement because of the lower specific energy per tonne of chlorine
- Dismantling and disposal of mercury-contaminated parts of the old plant.

Normally the decision for a conversion is initiated by plans for an expansion of the production capacity or by environmental legislation. Each change in the plant structure or in the cost structure may lead to reevaluation of the future of the electrolysis plant. Therefore, each plant has to be considered individually [190], [191]. For the European chlor-alkali industry a detailed analysis of the impact of a conversion of all mercury cells to the membrane technology on the competitiveness of the industry is given in [111].

The situation is different for diaphragm cell plants. These plants are still economic where inexpensive brine (e.g., from solution mining) is available, energy costs are comparably low (e.g., from cogeneration of electricity and steam on site), and when the market price for caustic is determined by the lower quality of diaphragm caustic. In countries like the United States it will be difficult to economically justify conversion. In contrast to the mercury process, improvements to the cells are still being made, resulting in lower operating costs and savings in solid waste disposal. The investment to convert a plant with a capacity of 1000 t/d to membrane technology is ca. 90 million dollars [187].

In the first few years after the introduction of membrane technology, diaphragm cells in several plants were equipped with membranes (retrofit) to reduce the cost of steam for cell liquor concentration, to give a small reduction in electricity consumption and better quality of caustic. This procedure is economic where steam is very expensive [188], [192].

9.3. Sodium Hydroxide and Potassium Hydroxide

9.3.1. Sodium Hydroxide [312]

NaOH is one of the oldest man-made chemicals. Excavations in Egypt show samples from 3000 B.C. Written records of the production by the reaction of sodium carbonate with calcined limestone came from Egypt and India, they are almost 2000 years old.

Physical and chemical properties. Solid sodium hydroxide has a white, crystalline appearance,

Molecular mass	39.997
Melting point	322 °C
Boiling point at 0.1 MPa	1388 °C
Heat of fusion	6.77 kJ/mol
Specific heat capacity at 20 °C	3.24 $Jkg^{-1}K^{-1}$
Density	1.77 g/cm^3

Solid NaOH is strongly hygroscopic, with water it forms six defined hydrates (see Fig. 30).

In presence of moist air, NaOH reacts with atmospheric carbon dioxide to form sodium carbonate. With moist carbon monoxide under pressure it forms sodium formate. Sodium hydroxide is a very effective drying agent. It is fairly soluble in methanol and ethanol. The heat of solution in water is ca 44 kJ/mol at 18 °C.

The densities of aqueous solutions at 20 °C are:

NaOH, wt%	20	30	40	50
Density, g/cm^3	1.219	1.328	1.430	1.525

The boiling point of the solution increases with concentration:

NaOH, wt%	5.9	23.1	33.8	48.3	54.6
b.p., °C	105	110	120	140	150

The viscosity of the 50% solution at 20 °C is 79 mPa.s.

Anhydrous NaOH reacts very slowly with most substances, at room temperature it attacks most metals only slowly (iron, magnesium, calcium). The corrosion rate increases rapidly with increasing temperature. Amphoteric metals such as zinc, aluminum, tin and lead are attacked by dilute NaOH solutions at room temperature, iron, stainless steel and nickel are fairly resistant.

Production of caustic soda solution. In 1998, the worldwide production capacity was about 54 million tons per year. Ca. 96–98% of this amount is produced by chloralkali electrolysis [313]. The three processes are described in detail in chapter 5 (Mercury Cell Process), chapter 6 (Diaphragm Process) and chapter 7 (Membrane process), a comparison of the relative qualities is given in chapter 9.

Other processes not linked to chlorine production are the old causticization of sodium carbonate:

$$Na_2CO_3 + CaO + H_2O \rightarrow 2\ NaOH + CaCO_3$$

This process is of economic interest whenever the demand for caustic soda exceeds the demand for chlorine and consequently the prices for NaOH are high. The source of sodium carbonate may be the Solvay process or the access to natural sodium carbonate or sodium hydrogen carbonate, e.g. trona ($Na_3H(CO_3)_2 \cdot 2\ H_2O$), a natural product which is used in Green River, Wyoming, United States.

Another chlorine-free route is the electrolytical or electro-dialytical splitting of sodium sulphate into sodium hydroxide and sulfuric acid. This process is of special

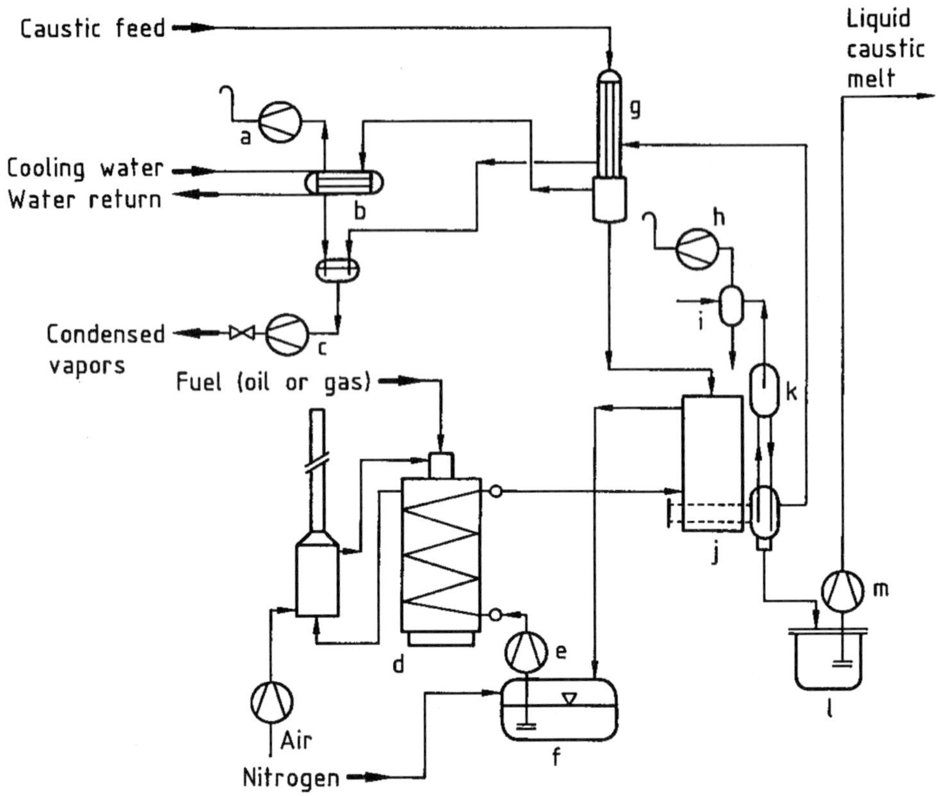

Figure 73. High-Concentration unit for caustic soda (Bertram process)
a) Vacuum pump; b) Condenser; c) Condensate pump; d) Furnace unit; e) Salt melt pump; f) Heated melt tank; g) Preconcentration unit; h) Vacuum pump; i) Condenser; j) Falling-film concentrator; k) Flash evaporator; l) Heated caustic melt tank; m) Caustic melt pump

interest in the pulp and paper industry in order to recycle the waste product sodium sulphate:

$$Na_2SO_4 + 2\,H_2O \rightarrow 2\,NaOH + H_2SO_4$$

Production of solid caustic soda. Usually, the electrolysis plants are producing a 50 % NaOH solution. Solid, 100 % NaOH is obtained by evaporating the solution until the water content is below 0.5 to 1.5 wt %.

In multistage processes, heat exchangers and falling film evaporators heat the solution up to > 400 °C (see Fig. 73).

The heat is generated electrically, a molten salt (an eutectic mixture of sodium and potassium nitrites and nitrates) is used as a heat transfer medium. The water vapors are removed by flashing. The product of the evaporation unit is a anhydrous melt, which is then cooled and transformed to the desired final product:

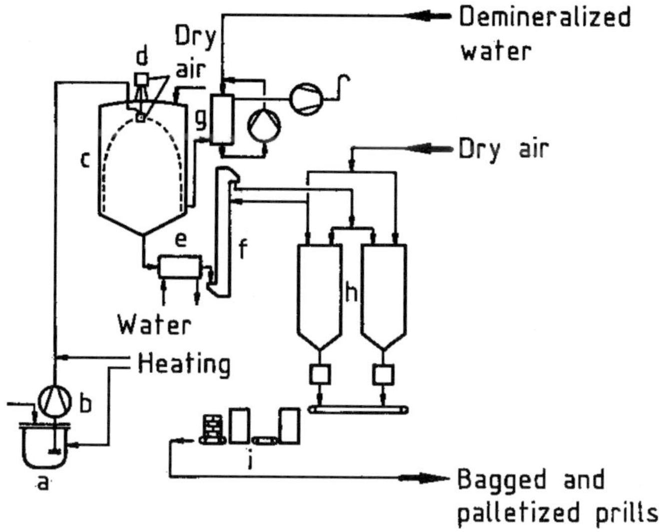

Figure 74. Production of sodium hydroxide prills (Bertrams process)
a) Caustic melt tank; b) Caustic melt pump; c) Spray tower; d) Rotating spray system; e) Product cooler; f) Elevator; g) Exhaust air treatment; h) Silo; i) Bagging and palletizing

Flakes are formed on a rotating drum with internal cooling, partially submersed in the melt. The layer of solidified NaOH is scrapped off from the drum by a knife and milled, giving the flakes, which are cooled further down and immediately packed in bags or drums.

Prills are produced by spraying the melt in prill towers. The droplets are cooled down to 250 °C during their free falling in cooled air. The uniform prills have diameters of 0.5 to 1.0 mm. After further cooling in a rotating cooling drum to 50 °C, they are stored in silos and packaged (Fig. 74).

Moulded pieces are obtained by pressing flakes or prills in briquetting presses which have a suitable formed surface.

Cast blocks: The molten NaOH is fed directly into steel drums, which are cooled by external water sprays to reduce the iron content of the product.

Storage, packaging and transportation. *Materials:* Pipelines, containers or storage tanks are made of carbon steel, stainless steel or rubberlined steel at temperatures < 50 °C. At higher temperatures austenitic Cr-Ni-steel, nickel, Hypalon, polypropylene, or epoxy resins are suitable. Heat exchangers are usually constructed of nickel, nickel alloys or stainless steel. Highly concentrated solutions at high temperatures e.g. in evaporating apparatus are made of nickel or are even silver clad. A review of materials for use with sodium hydroxide is given in [314].

Commonly, for storage and transportation a concentration of 50 % is preferred. Because of the freezing point of ca. 15 °C, the vessels must be insulated or be equipped with a heating device (e.g. vapor coil).

Packaging, transportation: 50 % sodium hydroxide solution is transferred over short distances by pipelines, otherwise by road or rail tank cars, by ships or barges, or in drums containing 100 to 250 kgs.

Solid caustic soda (prills, flakes) are stored in steel hoppers (30 – 50 m^3), sealed polyethylene bags (25 – 50 kg), big bags (up to 1000 kg), hoppers or containers (500 to 1000 kg and more).

The Chlorine Institute offers guidelines for storage equipment and piping systems [315], for the loading and unloading of tankcars and of tank motor vehicles [316], and the handling of caustic solution barges [317].

Labelling:	sodium hydroxide solution	sodium hydroxide solid
CAS-Nr:	[*1310-73-2*]	[*1310-73-2*]
Index-Nr:	011-002-00-6	011-002-00-6
EWG-Nr:	215-185-5	215-185-5
Symbol of danger:	C	C
R- and S-Phrases:	R 35 S 26, 36/37/39, 45	R 35 S 26, 37/39, 45
GGVE/GGVS/RID/ADR:	Cl. 8, Nr. 42b, Subs. Nr. 1824	Cl. 8, Nr. 41b, Subst. Nr. 1823
ADNR:	Class 8, Nr. 42b, 80/1824	Class 8, Nr. 41b, 80/1823
GGVSee/IMDG-Code:	Class 8, UN-Nr. 1824, PG II, EMS-Nr: 8-06, MFAG: 705	Class 8, UN-Nr. 1823, PG ll, EMS-Nr.8-06, MFAG: 705
ICAO/IATA-DGR:	Class 8, UN/ID-Nr. 1824, PG II	Class 8, UN/ID-Nr. 1823, PG ll

Safety precautions, First Aid: Caustic soda causes severe chemical burnings, (all contacts with sodium hydroxide during handling must be prevented).

In case of contact with eyes, rinse immediately with plenty of water and seek medical advice

Wear suitable protective clothing, gloves and eye/face protection (Protective goggles, safety gloves, if necessary: dust mask and protective clothing and shoes)

In case of accident or if you feel unwell, seek medical advice immediately

Quality, analysis: Standard methods for the chemical analysis of caustic soda and caustic potash are given in ASTM - E 291:

Total Alkalinity is determined by titration with standard hydrochloric acid solution using methyl orange indicator solution or modified methyl orange indicator solution

Sodium Carbonate or Potassium Carbonate: (Gas-volumetric method, the lower limit of determination is 0.001 g as carbon dioxide): Carbon dioxide is evolved by acid decomposition of carbonate in the sample. The volume of CO_2 is measured and calculated as sodium or potassium carbonate.

(Gravimetric method, the lower limit of determination is 0.001 g as carbon dioxide): Carbon dioxide is evolved by acid decomposition of the carbonate in the sample and is

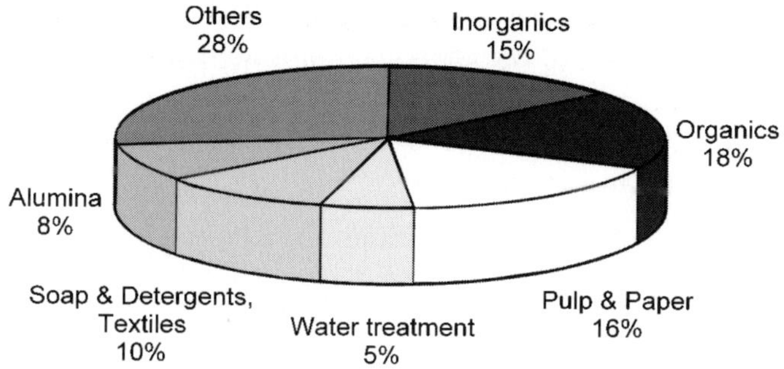

Figure 75. Uses of caustic soda, world 1997

absorbed on sodium-asbestos absorbent. The increase in weight is a measure of the carbonate present.

Chloride, Titrimetric (the lower limit of determination is 0.001 g as chloride): The sample is diluted, acidified, and treated with a small excess of standard silver nitrate solution. The precipitated silver chloride is removed by filtration and the excess silver nitrate is titrated with standard ammonium thiocyanate solution using ferric ammonium sulfate indicator.

Chloride, Ion Sensitive Electrode (determination of 0.6 ppm to 120 ppm chloride): The sample is acidified, followed by immersion of a chloride ion selective electrode into the sample solution and measurement of the mV response. Comparison of the response to a standard calibration curve allows interpolation of chloride amount.

Iron (the lower limit of this photometric determination is 0.1 ppm as Fe): Iron is reduced to the ferrous condition where it forms an orange-red complex with 1,10-phenanthroline (orthophenanthroline) in an acetate-buffered solution at pH 5. Intensity of the color so formed is measured at 510 nm in a photometer calibrated with standard iron solutions. The color develops within 15 min, is very stable, and follows Beer's law.

Sulfate (the lower limit of determination is 0.002 g as SO_3): Sulfate is determined gravically by precipitation as barium sulfate which is filtered off, washed, ignited, and weighed.

Uses. The world's end use sectors for caustic soda in 1997 are shown in Fig. 75 [318]. In Europe, the percentages are comparable. In regions like Australia, Surinam or Venezuela, with a strong aluminum industry or in Scandinavia, Canada with pulp and paper industry the relativ applications are predominant.

The world caustic soda demand was 44.3 million tonnes in 1997.

Use of caustic soda in the world 1997:

Inorg. Chemicals	15 %
Organic Chemicals	18 %
Pulp & Paper	16 %
Water Treatment	5 %
Soap & Detergents, Textiles	10 %
Alumina	8 %
Others	28 %

Between 42 and 50% of the production is used in the chemical industry. Most customers require dilute aqueous solutions, the application of waterfree solid caustic soda is limited to drying processes, waterfree organic reactions, and to exports into countries with logistic problems in handling greater amounts of liquid solutions; the share is only 2–3% of the production.

- The organic chemical industry uses NaOH for saponification reactions, production of anionic intermediates, etherification and esterification, basic catalysis, for waste gas scrubbing and for waste water neutralisation.
- The inorganic chemistry uses NaOH in the manufacture of sodium salts, for alkaline ore digestion and for pH-adjustment.
- The pulp & paper industry uses NaOH for the extraction of lignin during pulp leaching.
- The textile industry uses NaOH in the manufacture of viscose and of cellophane. It was this application which promoted the development of the mercury cell because this process produces the chloride-free "rayon quality" caustic soda. Cotton and wool can be improved by mercerization, a treatment with NaOH.
- The soap & detergents industry uses NaOH for sodium phosphate production, soap manufacture by saponification of fats and oils, and the production of detergents from organic sulfonic acid and NaOH.
- The aluminum industry uses NaOH in the Bayer-Process for treating bauxite.
- In the water treatment sector, NaOH is used for the regeneration of ion exchangers for water purification, pH-adjustment, waste water treatment, drinking and swimming pool water disinfection by treatment with hypochlorite solution.
- Other uses of NaOH include electroplating technology, treatment of natural gas (removal of hydrogen sulfide), petroleum and refining industry, steel industry, gold extraction by cyanide leaching, food industry (e.g. cleansing of apparatus), manufacture of soda ash in competition to the Solvay process and others.

World trade: Most chlor-alkali electrolysis plants are situated in the vicinity of the chlorine users. The alkali users, particularly the alumina and pulp & paper industries, are placed in the winning regions. Like the world trade with chlorinated hydrocarbons there are flows of liquid caustic soda with a volume between 2 and 3 million tonnes of NaOH (100%) per year. The net exporting regions are and will remain North America, the Middle East, Japan and Western Europe. Importing regions are South America (Surinam, Venezuela), Australia and South East Asia.

The cycles of economy are temporarily different for chlorine derivatives and caustic soda. This fact creates imbalances in the demand for chlorine and caustic and as a consequence strongly varying prices: between 1987 and 1997, the contract prices for caustic from the US Gulf Coast to Australia fluctuated from ca. 40 US$ to 290 US$/tonne [319].

9.3.2. Potassium Hydroxide [318a]

Properties. Pure, solid potassium hydroxide [*1310-58-3*], KOH, caustic potash, M_r 56.11, ϱ 2.044 g/cm^3, mp 410 °C, bp 1327 °C, heat of fusion 7.5 kJ/mol, is a hard, white substance. It is deliquescent and absorbs water vapor and carbon dioxide from the air. Potassium hydroxide dissolves readily in alcohols and water (heat of solution 53.51 kJ/mol). The solubility of KOH (g KOH/100 g H_2O) in water is shown below:

Temperature, °C	0	10	20	30	50	100
Solubility	97	103	112	126	140	178

The mono-, di-, and tetrahydrates are formed with water. Aqueous potassium hydroxide is a colorless, strongly basic, soapy, caustic liquid, whose density depends on the concentration:

Concentration, wt %	10	20	30	40	50
Density, g/cm^3	1.092	1.188	1.291	1.395	1.514

Technical caustic potash (90 – 92 % KOH) melts at ca. 250 °C; the heat of fusion is ca. 6.7 kJ/mol.

Production. Today, potassium hydroxide is manufactured almost exclusively by potassium chloride electrolysis. The diaphragm, mercury, and membrane processes are all suitable for the production of potassium hydroxide, but the mercury process is preferred because it yields a chemically pure 50 % potassium hydroxide solution.

In the *diaphragm process*, a KCl-containing, 8 – 10 % potassium hydroxide solution is initially formed, whose salt content can be reduced to ca. 1.0 – 1.5 % KCl by evaporation to a 50% liquor. Further purification is complicated, and the quality of liquor from mercury cells cannot be achieved.

In the *mercury process* a very pure KCl brine must be utilized, because even traces (ppb range) of heavy metals such as chromium, tungsten, molybdenum, and vanadium, as well as small amounts (ppm range) of calcium or magnesium, lead to strong evolution of hydrogen at the amalgam cathode. The very pure potassium hydroxide solution running off the decomposers is cooled, freed from small amounts of mercury in precoated filters, and in some cases sent immediately to the consumer as a 45 – 50 % liquor in drums, tank cars, or barges.

Since about 1985, new cell rooms for the manufacture of potassium hydroxide solution have used the *membrane process*. At present, the cell liquor has a low chloride content (10 – 50 ppm); the KOH concentration is 32%. Before dispatch, it is concentrated to 45 – 50 % by evaporation.

Nonelectrochemical processes have been proposed for the manufacture of chlorine and potassium hydroxide from KCl by thermal decomposition of potassium nitrite in the presence of Fe_2O_3 [399].

This method involves reacting KCl with NO_2 to obtain Cl_2 and potassium nitrite, reacting the KNO_2 with iron(III) oxide and oxygen to give potassium ferrate ($K_2Fe_2O_4$), and reacting the ferrate with water to produce KOH. Another method consists of reacting an aqueous solution of KCl with NO_2 and O_2 to give Cl_2 and KNO_3, which is reacted with water in the presence of Fe_2O_3 to produce KOH.

Largely water-free, ca. 90 – 95 % potassium hydroxide (caustic potash) is obtained by evaporating potassium hydroxide solution. The residual content of 5 – 10 % H_2O is bound as a monohydrate.

Suitable evaporation processes are single- or multistage falling-film evaporators [400], Badger single-tube evaporators, or boilers connected in cascade. Heating is carried out with steam or by means of heat-transfer agents (salt melts, Dow-therm). Flash evaporators are used as the final stage in large-capacity plants [401].

To counter the strong corrosiveness of the potassium hydroxide solution and retain the purity of the caustic potash, the equipment is made largely from high-purity nickel (LC 99.2) or is silverplated. The equipment is often protected by polarization.

For dispatch, caustic potash comes on the market poured directly into drums or packed in polyethylene bags after cooling; in blocks, molded pieces, flakes, prills, and as a powder. Potassium hydroxide is classified as a corrosive material:

UN no.	1814	(for aqueous solution)
UN no.	1813	(for dry material)
GGVS/GGVE	Class 8	
RID/ADR	Class 8	

Handling is described in [402].

Quality Specifications, Analytics. Potassium hydroxide solution is supplied in pure quality [total alkalinity 49.7 – 50.3 %, KOH 48.8 % (min.), NaOH 0.5 % (max.), CO_3^{2-} 0.1 % (max.)] or in technical quality [total alkalinity 49.7 – 50.3 %, NaOH 1.0 % (max.), CO_3^{2-} 0.3 % (max.)]. The contents of Cl^-, SO_4^{2-}, Fe^{2+}, and Ca^{2+} are < 30 ppm. Solid caustic potash produced from amalgam liquor has a total alkalinity (calculated as KOH) of 89 – 92 %, NaOH 1.5 % (max.), CO_3^{2-} 0.5 % (max.), Cl^- 0.01 % (max.). The values for SO_4^{2-}, Fe^{2+}, and Ni^{2+} are < 50 ppm. Caustic potash from diaphragm electrolysis has a Cl^- content of 2.5 – 3.0 % and higher content of heavy metals.

Analysis. The *total alkalinity* includes $K_2CO_3 + KOH + Na_2CO_3$; it is determined with 0.5 N H_2SO_4 by potentiometric titration or with a methyl orange indicator (change to brown-red). Sodium is determined by flame photometry. The *chloride content* is determined by turbidity measurement after addition of $AgNO_3$. The *sulfate content* is determined by ion chromatography or gravimetrically after precipitation as barium sulfate. The *metal content* is determined by atomic absorption spectroscopy or photometrically by complex formation (Fe^{2+} as sulfosalicylate, Si^{4+} as the molybdato complex, Cu^{2+} as pyrrolidinodithiocarbamate, and Ni^{2+} as the diacetylglyoxime complex).

Test methods for *photographic-grade* potassium carbonate, anhydrous are described in ISO 3623-1976 (E).

Economic Aspects and Uses. Pure-quality potassium hydroxide is used as a raw material for the chemical and pharmaceutical industry, in dye synthesis, for photography as a developer alkali, and as an electrolyte in batteries and in the electrolysis of water. Technical-quality KOH is used as a raw material in the detergent and soap industry; as a starting material for inorganic and organic potassium compounds and salts (e.g., potassium carbonate, phosphates, silicate, permanganate, cyanide); for the manufacture of cosmetics, glass, and textiles; for desulfurizing crude oil; as a drying agent; and as an absorbent for carbon dioxide and nitrogen oxides from gases.

World production is estimated at ca. $700-800 \times 10^3$ t/a. Main producers are the United States [403], Germany, Japan, and France. Other important producer countries are Belgium, the United Kingdom, Italy, Spain, South Korea, India, Israel, Jugoslavia, Czechoslovakia, Sweden, and Romania.

10. Other Production Processes

10.1. Electrolysis of Hydrochloric Acid

Electrolytic decomposition of aqueous hydrochloric acid is used to produce chlorine and hydrogen. The first pilot plant was set up by G. MESSNER in 1942 in Bitterfeld, Germany, and since 1964 eight full-scale plants have been commissioned in Europe and the United States, a total capacity of 540 000 t/a [193]. Hydrogen chloride is a byproduct of many organic industrial processes. Electrolysis of hydrochloric acid competes with chemical processes in which either hydrogen chloride is used to produce chlorinated hydrocarbons directly, e.g., by oxychlorination, or where chlorine is produced by chemical reaction, e.g., in the KEL chlorine process (see p. 136). The advantages of the electrolytic process are very pure products without further treatment, reliability (simple design), ease of operation, flexibility (5 : 1 turndown ratio), and low energy consumption even with small installations.

Principles. Hydrochloric acid (22 wt % HCl) is fed into the cells in two separate circuits, a catholyte circuit and an anolyte circuit. During electrolysis the concentration is reduced to ca. 17 %, and the temperature increases from 65 to 80 °C. A part of the depleted acid is separated from the catholyte stream, concentrated in the absorption plant to ca. 30 %, and fed back into the main stream. The electrolyzer is bipolar, with pairs of electrodes arranged like the leaves of a filter press. A diaphragm or membrane (e.g., Nafion 430) separates the anode compartment from the cathode compartment to prevent mixing of the gaseous products.

The reversible standard decomposition potential of hydrochloric acid is 1.358 V, made up of the anode potential, the discharge of chloride ions with formation of chlorine, and the cathode potential, the discharge of hydroxonium (H_3O^+) ions with formation of hydrogen. In practice (> 15 % HCl, 70 °C), the decomposition potential is $\leq$ 1.16 V.

The graphite electrode plates are not attacked by 22 % hydrochloric acid. A poly(vinyl chloride) (PVC) fabric constitutes the diaphragm. Chlorine dissolved in the anolyte diffuses through the diaphragm and is reduced at the cathode, causing a loss of 2 – 2.5 % of the theoretical current yield. The increase of cell voltage when current flows is mainly because of the hydrogen overpotential at the graphite cathode and the resistance of the electrolyte. Depolarizing agents (polyvalent metal ions) in the catholyte reduce the overpotential by $\leq$ 300 mV at 4 kA/m^2 [194].

The conductivity of hydrochloric acid is maximized at a concentration of 18.5 wt %. High temperatures improve the conductivity, but to avoid increased vapor pressure of HCl and material problems, the temperature is kept below 85 °C. Modern cells have a voltage of ca. 1.90 V at 4.8 kA/m^2, corresponding to an energy consumption of 1400 – 1500 kW h per tonne of chlorine.

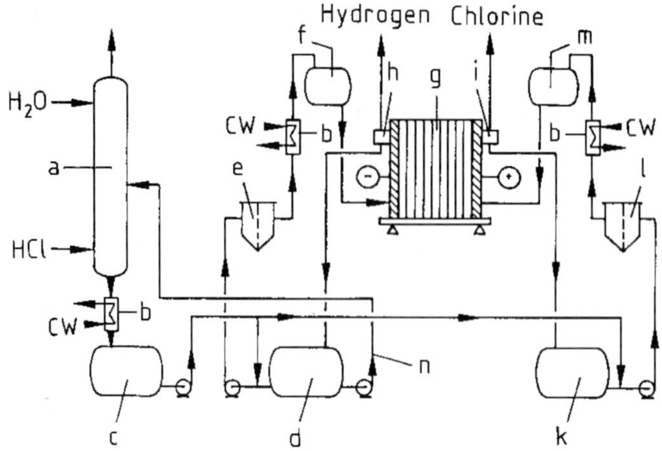

Figure 76. Simplified flow diagram of a hydrochloric acid electrolysis
a) Absorption column; b) Heat exchanger; c) Strong acid tank; d) Catholyte collecting tank; e) Catholyte filter; f) Catholyte supply tank; g) Electrolyzer; h) Hydrogen – catholyte separator; i) Chlorine – anolyte separator; k) Anolyte collecting tank; l) Anolyte filter; m) Anolyte supply tank; n) Weak acid line to absorber

Diaphragm Cells. Hydrochloric acid electrolysis cells are manufactured by Hoechst – Uhde [195]. Each Hoechst – Uhde electrolyzer consists of 30 – 36 individual cells that are formed from vertical graphite plates connected in series, between which there are diaphragms. To improve gas release, vertical slits are milled in the graphite plates, which are cemented in frames made of HCl-resistant plastics. At the bottom of the frames, channels feed in the electrolyte. The gases rise up the plates and pass through ducts into collection channels in the upper part of the cell. Chlorine leaves the cell with the anolyte, and hydrogen leaves with the catholyte. The end plates of the electrolyzer are made of steel lined internally with rubber and are held together by spring-loaded tension rods. The electric current is supplied via graphite terminals. The unit rests on insulated steel frames. The effective surface of the electrodes is 2.5 m^2, and the current loading can be up to 12 kA.

DeNora and General Electric are developing an electrolyzer with a solid polymer electrolyte (SPE) based on Nafion [196]. In addition to a voltage savings of 20 %, it is hoped that completely chloride-free hydrogen gas can be produced.

Operation. A simplified flow diagram of the process as operated by Bayer – Hoechst – Uhde is shown in Figure 76.

In the absorption column, the hydrogen chloride gas is absorbed adiabatically by depleted hydrochloric acid from the catholyte. In the upper section of the column, an absorber removes the remaining hydrogen chloride and the water vapor by absorption in a water stream, which makes up the water balance of the process. The 30 wt% acid that is produced is then cooled, purified if necessary by activated carbon, and supplied to the anolyte and catholyte circulation systems.

The electrolyte is pumped through a filter and heat exchanger to a gravity feed tank for the electrolyzer unit. The gases produced are freed from the electrolytes in separators, and the electrolytes flow back into their respective collecting tanks to be resaturated. The working life of the PVC diaphragms, 1–2 years, depends on the impurities in the acid. The concentrated acid is, therefore, purified carefully [197].

The product gases are saturated with water vapor and hydrogen chloride at the partial vapor pressures of 20% hydrochloric acid. Both product streams are cooled. Sodium hydroxide solution is used to wash the hydrogen, removing chlorine and hydrogen chloride and producing a 99.9% product. The chlorine, which is dried by sulfuric acid, contains ca. 0.5% hydrogen and ca. 0.05% carbon dioxide. The hydrogen overpotential can be reduced by activation of the cathodes.

Membrane Cells [198], [199]. Since 1992 Bayer has replaced the woven fabric cloth in the diaphragm cells by anion-exchange membranes of the sulfonate type. Only hydrated protons are able to pass from the anolyte to the catholyte, so that the whole cell and the electrolyte systems are simplified. Together with an optimized surface of the electrodes for better gas release, this leads to:

- Lower cell voltage of 300 mV, corresponding to a power consumption of 1300 kWh per tonne of chlorine at 4.8 kA/m^2
- Longer life of the cell components
- Higher product quality
- Improved safety of operation
- Simplified process

A similar electrolytic process for recovering chlorine from anhydrous HCl, also using membrane cell technology, has been developed by DuPont [200].

10.2. Chemical Processes

The chlor-alkali process produces chlorine and sodium hydroxide solution in fixed stoichiometric proportions. Experience has shown that there tends to be a surplus of either chlorine or sodium hydroxide. Chlorine may, however, be produced competitively without the byproduct sodium hydroxide by nonelectrolytic methods. The starting material is usually hydrogen chloride, which is catalytically oxidized to chlorine by oxygen, air, nitric acid, sulfur trioxide, or hydrogen peroxide. Other processes start from ammonium chloride or metal chlorides.

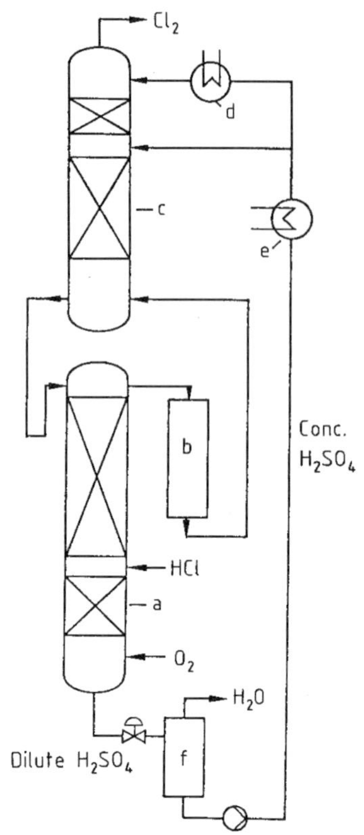

Figure 77. Flow diagram of the KEL chlorine process (simplified)
a) Stripper; b) Oxidizer; c) Absorber – oxidizer; d) Acid chiller; e) Acid cooler; f) Vacuum flash evaporator

10.2.1. Catalytic Oxidation of Hydrogen Chloride by Oxygen

A catalyst is essential for the economic oxidation of hydrogen chloride to chlorine by air or oxygen (Deacon Process), and the catalyst must be active at low temperature and have adequate life. There are many patents claiming improved catalysts and equipment. Most of the catalysts are oxides and/or chlorides of metals on various substrates. Only three processes have been commercialized.

The KEL Chlorine Process. The process developed by KELLOGG [198] uses concentrated sulfuric acid (ca. 80 %) with ca. 1 % nitrosylsulfuric acid as the catalyst. From 1975 to 1988 Du Pont operated a full-scale plant in Corpus Christi, Texas, recovering up to 600 t/d of chlorine. The plant was shut down due to a change in the structure of the plant and because of material problems after more than 10 years of operation. The raw material, from a fluorinated hydrocarbon plant, consisted of waste gases that contained hydrogen chloride [201]. Figure 77 shows a simplified flow diagram.

Sulfuric acid catalyst is fed into the top of the stripper column. The hydrogen chloride gas reacts with the catalyst to form nitrosyl chloride:

$$HCl + NOHSO_4 \longrightarrow NOCl + H_2SO_4$$

The oxygen, the ultimate oxidizing agent, blows the remaining hydrogen chloride out of the sulfuric acid, which becomes more concentrated and also is cooled in a flash vaporizer. This acid is then fed back into the process. Nitrosyl chloride, hydrogen chloride, oxygen, and water vapor flow as a gaseous stream into the oxidizer and react there, increasing the temperature:

$$2\ NOCl + O_2 \longrightarrow 2\ NO_2 + Cl_2$$

$$NO_2 + 2\ HCl \longrightarrow NO + Cl_2 + H_2O$$

In the absorber – oxidizer, the rest of the hydrogen chloride is oxidized. Concentrated sulfuric acid is fed in at the top, reacts with the oxides of nitrogen to form nitrosylsulfuric acid, absorbs the water that has formed, and is conducted back into the stripper:

$$NO + NO_2 + 2\ H_2SO_4 \longrightarrow 2\ NOHSO_4 + H_2O$$

$$NOCl + H_2SO_4 \longrightarrow NOHSO_4 + HCl$$

The cooled, dried chlorine gas still contains ca. 2 % hydrogen chloride and up to 10 % oxygen. Both are removed by liquefaction.

The net reaction is

$$4\ HCl + O_2 \longrightarrow 2\ Cl_2 + 2\ H_2O$$

The installation at Corpus Christi operated at 1.4 MPa and 120 – 180 °C. On account of the aggressive nature of the chemicals, expensive materials, such as tantalum-plated equipment and pipes, must be used. For outputs of 250 – 300 t of chlorine per day, this process can be more economical than the electrolysis of hydrochloric acid, depending on local conditions.

The Shell Chlorine Process. The catalyst developed by Shell consists of a mixture of copper(II) chloride and other metallic chlorides on a silicate carrier [202]. The reaction of the stoichiometric mixture of hydrogen chloride and air takes place in a fluidized-bed reactor at ca. 365 °C and 0.1 – 0.2 MPa. The yield is 75 %. The water condenses out from the gas stream, and the hydrogen chloride is removed by washing with dilute hydrochloric acid. After the residual gas has been dried with concentrated sulfuric acid, the chlorine is selectively absorbed, e.g., by disulfur dichloride. After desorption and liquefaction, the chlorine has a purity > 99.95 %.

A manufacturing unit was built by Shell in the Netherlands, 41 000 t/a, and another in India, 27 000 t/a, but both have been closed down owing to the prolonged surplus of chlorine on the market.

The Mitsui MT-Chlorine Process. The catalyst consists of chromium(III) oxide on a silicate carrier [203]. In a fluidized-bed reactor, hydrogen chloride is reacted with oxygen at a temperature of 415 °C to give chlorine gas with a conversion rate of 73 – 77 %. The reactor is made from nickel-lined low carbon steel. The concentration of the purified product is > 99.5 % Cl_2.

A commercial plant for 30 000 t Cl_2/a is successfully operating since 1988, with an expansion to 60 000 t Cl_2/a in 1990.

A two-stage cyclic fluidized bed process for converting HCl to chlorine is described in [204]. The catalytic oxidation process combines the exothermic oxidation of 60 – 70 % of the HCl at 380 – 400 °C in a fluidized bed of copper oxychlorides impregnated on zeolite with the transfer of the reaction products to a second reactor operating at 180 – 200 °C where the rest of HCl is converted.

10.2.2. Oxidation of Hydrogen Chloride by Nitric Acid

The nitrosyl chloride route to chlorine is based on the strongly oxidizing properties of nitric acid:

$$6\ HCl + 2\ HNO_3 \longrightarrow 2\ Cl_2 + 2\ NOCl + 4\ H_2O$$

$$2\ NOCl + 2\ H_2O + O_2 \longrightarrow 2\ HCl + 2\ HNO_3$$

The practical problems lie in the separation of the chlorine from the hydrogen chloride and nitrous gases. The dilute nitric acid must be reconcentrated. Corrosion problems are severe. Suggested improvements include (1) oxidation of concentrated solutions of chlorides, e.g., LiCl, by nitrates followed by separation of chlorine from nitrosyl chloride by distillation at 135 °C or (2) oxidation by a mixture of nitric and sulfuric acids with separation of the product chlorine and nitrogen dioxide by liquefaction and fractional distillation [205].

10.2.3. Production of Chlorine from Chlorides

Alkali-metal chlorides, ammonium chloride, and other metallic chlorides are reacted, usually with nitric acid, to produce nitrate fertilizers [206]. Chlorine is not produced directly, but it can be obtained from the intermediate products nitrosyl chloride or hydrochloric acid.

11. Chlorine Purification and Liquefaction

Chlorine produced by the various processes, especially by electrolysis, is saturated with water vapor at high temperature and may also contain brine mist and traces of chlorinated hydrocarbons, and is normally at atmospheric pressure. Before the chlorine can be used, it must be cooled, dried, purified, compressed, and where necessary, liquefied. A simplified flow sheet is shown in Figure 78.

11.1. Cooling

Table 22 shows the volume, water content, and heat content of 1 kg of chlorine gas at 101.3 kPa as a function of temperature. To avoid solid chlorine hydrate formation, the gas is not cooled below 10 °C [207]. Cooling is accomplished in either one stage with chilled water or in two stages with chilled water only in the second stage.

The chlorine gas can be *cooled indirectly* in a tubular titanium heat exchanger so that the cooling water is not contaminated and the pressure drop is small. The resultant condensate is either fed back into the brine system of the mercury process or dechlorinated by evaporation in the case of the diaphragm process.

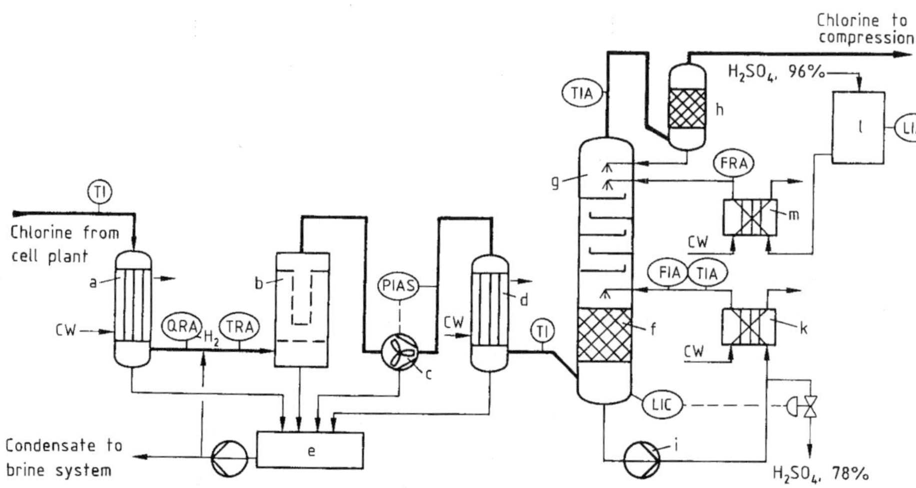

Figure 78. Simplified flow diagram of a chlorine processing plant
a) Chlorine gas cooler (primary); b) Chlorine demister; c) Blower or fan; d) Chlorine gas cooler/chiller (secondary); e) Condensate collection tank; f) Drier, first stage; g) Drier, second stage; h) Sulfuric acid mist separator; i) Sulfuric acid circulation pump; k) Cooler for circulating sulfuric acid; l) Sulfuric acid feed tank; m) Cooler for sulfuric acid feed

Table 22. Volume, moisture content, and enthalpy of 1 kg of chlorine gas at 101.3 kPa as a function of temperature *t*

t, °C	Volume, m^3		Water content, g/kg **	Heat content, kJ/kg *
	Dry	Saturated *		
0	0.312	0.314	1.54	3.81
20	0.335	0.342	5.95	24.45
40	0.357	0.385	19.7	69.50
60	0.380	0.473	61.5	188.41
70	0.392	0.565	112.0	325.73
80	0.404	0.756	222.0	623.83

* Chlorine gas saturated with water vapor at temperature *t*.
** Grams of H_2O per kg of Cl_2.

The chlorine gas can be *cooled directly* in packed towers. Water is sprayed into the top and flows countercurrent to the chlorine. This treatment thoroughly washes the chlorine; however, dechlorination of the wastewater consumes a large amount of energy. The cooling water should be free of traces of ammonium salts to avoid the formation of nitrogen trichloride.

Closed-circuit direct cooling of chlorine combines the advantages of the two methods. The chlorine-laden water from the cooling tower is cooled in titanium plate coolers and recycled. The surplus condensate is treated like the condensate from indirect cooling. Spray towers, as well as packed towers, are used. Water carry-over is removed by demisters, which reduce the amount of sulfuric acid used for drying.

11.2. Chlorine Purification

Water droplets and impurities such as brine mist are mechanically removed by special filter elements with glass wool fillings. The efficiency varies with the gas throughput. A commonly used device is the Brink *demister* [208]. Instead of glass wool, porous quartz granules can be used.

In *electrostatic purification,* the wet chlorine gas is passed between wire electrodes in vertical tubes. The electrodes are maintained at a d.c. potential of 50 kV with a current density of 0.2 mA/m^2. The particles and droplets in the chlorine become charged and collect on the tube walls. The resultant liquid is fed back into the brine system or chemically treated before disposal.

Activated carbon filters can adsorb organic impurities and may be regenerated by heating to 200 °C.

Gaseous impurities can be removed by *absorption* of the chlorine in a suitable solvent, such as carbon tetrachloride, water, or disulfur dichloride, followed by *desorption.* This can be coupled with further processes, such as the recovery of chlorine from the waste gas remaining after liquefaction [15, pp. 418–422].

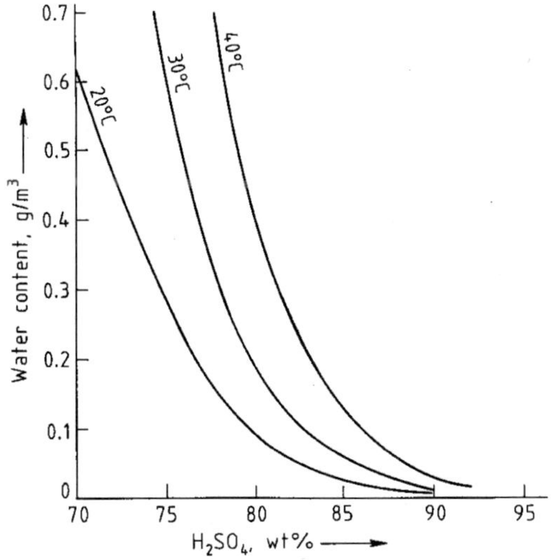

Figure 79. Drying chlorine with sulfuric acid Attainable moisture content as a function of concentration and temperature of the acid

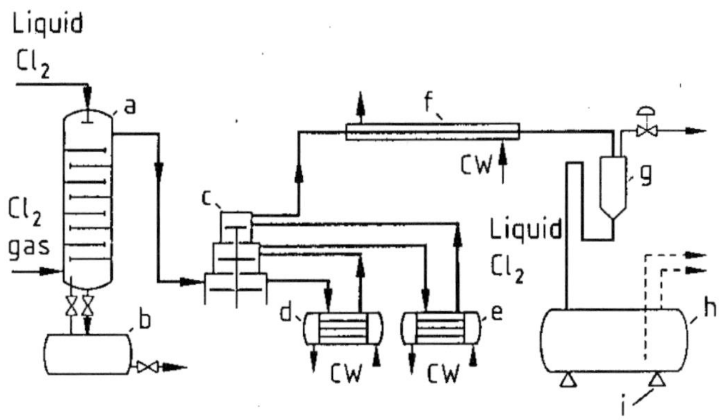

Figure 80. Multistage reciprocating compressor for chlorine liquefaction at 1 MPa with cooling water at 15 °C with liquid chlorine scrubbing
a) Low-temperature cooling and scrubbing column; b) Collection tank for impurities; c) Three-stage compressor; d) Intermediate cooler, stage 1; e) Intermediate cooler, stage 2; f) Liquefier; g) Chlorine collection vessel; h) Chlorine storage tank; i) Chlorine storage tank on load cells

A wash with concentrated hydrochloric acid removes the dangerously explosive nitrogen trichloride [209]. *Scrubbing with liquid chlorine* (see Fig. 80) mainly reduces the content of organic impurities and carbon dioxide, but it can also lower the bromine content. When the chlorine is cooled down to near its dew point, liquid chlorine scrubbing is often combined with compression by turbo or reciprocating compressors.

11.3. Drying

Drying of chlorine is carried out almost exclusively with concentrated sulfuric acid (96 – 98 wt%) [210]. Depending on the desired final concentration of the waste acid, drying can be a two-, three-, or four-stage process. The acid and chlorine flow countercurrently. The final moisture content depends on the concentration and temperature of the acid in the final stage (Fig. 79). An upper limit is 50 ppm H_2O. Low-temperature liquefaction (– 70 °C) demands lower moisture content, which can be achieved with molecular sieves, whereby 2.5 ppm is possible [211].

The *packed towers* usual in the first stages are constructed of rubber-lined steel or glass-fiber-reinforced poly(vinyl chloride). The heat liberated on dilution of the circulating acid is removed by titanium heat exchangers, and the weak acid is dechlorinated chemically or by blowing air. Often the acid is recirculated after reconcentration to 96 % by heating under vacuum. Generally, columns with *bubble cap plates* or *sieve trays* are used at the final stage. The drying is effective, but the pressure drop is great. Occasionally, spray towers are used to dry chlorine.

After drying, the chlorine gas is passed through a demister or a packed bed to remove sulfuric acid mist.

11.4. Transfer and Compression

In all operations involving compression, care must be exercised to prevent the heat of compression from increasing the temperature enough to ignite material in contact with the chlorine.

Wet chlorine gas can be compressed 20 – 50 kPa by a single-stage *blower* or *fan* with a rubber-lined steel casing and titanium impeller. It can also be compressed in *liquid-ring compressors,* so that further treatment of the chlorine can be accomplished in smaller equipment [212]. Sulfuric acid ring compressors are used for throughputs of 150 t of dry chlorine gas per day per compressor and for pressures of 0.4 MPa or, in two-stage compressors, 1.2 MPa. The heat of compression is removed by cooling the circulating liquid; cooling of the gas is not necessary. Advantages are simplicity of construction, strength, and reliability, but efficiency is low [213].

Reciprocating compressors were formerly lubricated with sulfuric acid, but are now available as *dry-ring compressors* (no lubrication). They can compress up to 200 t per day. Multistage compressors produce pressures up to 1.6 MPa. The heat of compression of each stage must be removed by heat exchangers or by injection of liquid chlorine (see Fig. 80). Well-purified chlorine gas is essential for trouble-free operation [214].

Turbo compressors are most economical when they operate with large amounts of chlorine. Each unit compresses up to 1800 t/d. In multiple-stage operation, pressures up to 1.6 MPa are reached. Labyrinth seals are used on the high-speed shafts. Require-

Table 23. Electrical energy requirement for compression and liquefaction of 1 t of chlorine gas

	Liquefaction pressure, MPa			
	0.1	0.3	0.8	1.6
Energy for compression, kW h/t	5	23	42	57
Energy for cooling, kW h/t	87	68	27	3
Combined energy, kW h/t	*92*	*91*	*69*	*60*
Starting temperature, °C	–36	–8	25	53
Final temperature, °C	–42	–17	14	40

ments for cooling and gas purity are like those of reciprocating compressors. *Screw compressors* handle low rates of chlorine and give pressures up to 0.6 MPa. *Sundyne blowers* are one-stage high-speed centrifugal compressors handling 80 – 250 t per day and giving pressures up to 0.3 MPa. Liquid chlorine injection is used for cooling [215]. *Membrane compressors* are used for pressurizing storage tanks with chlorine gas to transfer liquid chlorine to other vessels [216]. Liquid chlorine is pumped with *canned pumps* [217].

11.5. Liquefaction

The most suitable liquefaction conditions can be selected within wide limits. Important factors are the composition of the chlorine gas, the desired purity of the liquid chlorine, and the desired yield. There are nomograms that give the relationship between the chlorine concentrations of the incoming and residual gases, liquefaction yields, pressures, and temperatures [218]. Increasing the liquefaction pressure increases the energy cost of chlorine compression, although the necessary amount of cooling decreases, resulting in an overall reduction in energy requirement (Table 23) [219].

Any hydrogen is concentrated in the residual gas. To keep the hydrogen concentration below the 6 % explosive limit, conversion of gas to liquid should be limited to 90 – 95 % in a single-stage installation. Higher yields may be obtained by condensing the chlorine from the residual gas in a second stage, which is constructed to reduce the risk from explosion [220]. This is achieved by the use of sufficiently strong equipment to withstand explosions or by the addition of enough inert gas to keep the mixture below the explosive limit. Multistage installations can liquefy over 99.8 % of the chlorine gas.

High-pressure (0.7 – 1.6 MPa) liquefaction with *water cooling* (Fig. 80) does not require a cooling plant. Therefore, it has the lowest energy cost of all methods; however, the high construction cost must be set against this.

Medium-pressure (0.2 – 0.6 MPa) liquefaction with *cooling* (– 10 to – 20 °C) is especially useful when only a part of the chlorine is to be liquefied and the remaining gas is to be reacted at the liquefaction pressure, e.g., with ethylene to form ethylene dichloride. The residual gas can be fed into the compressor suction systems, provided that the

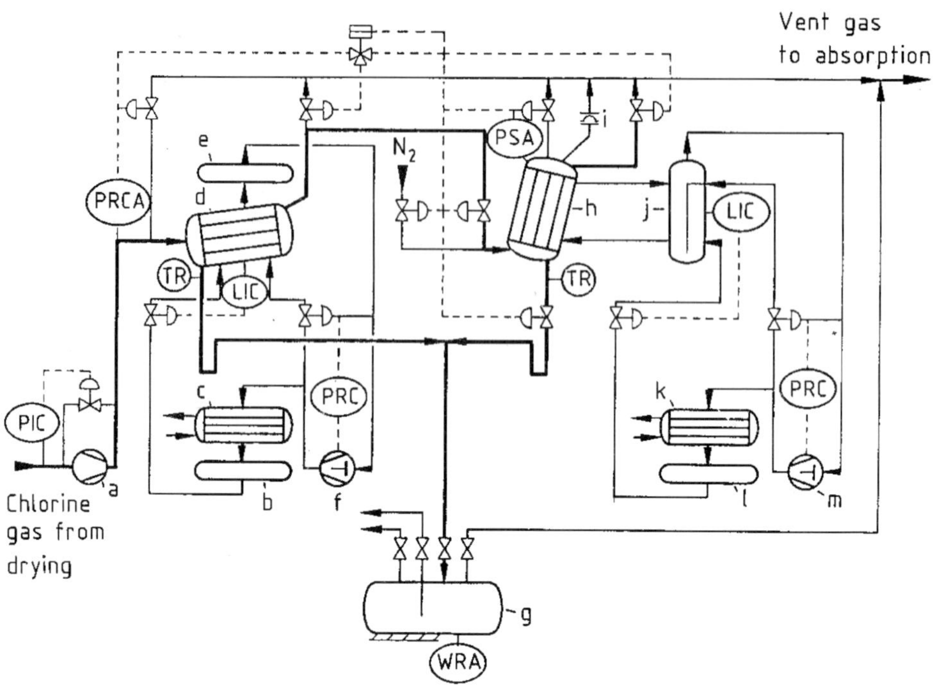

Figure 81. Flow diagram of a two-stage chlorine liquefaction plant at intermediate pressure — Uhde system
a) Chlorine gas compressor; b) Refrigerant collector, stage 1; c) Refrigerant condenser, stage 1; d) Chlorine liquefier, stage 1; e) Refrigerant separator, stage 1; f) Refrigerant compressor, stage 1; g) Liquid chlorine storage tank; h) Chlorine liquefier, stage 2; i) Bursting disk; j) Refrigerant separator, stage 2; k) Refrigerant condenser, stage 2; l) Refrigerant collector, stage 2; m) Refrigerant compressor, stage 2

increased inert gas content does not interfere with the subsequent process. Otherwise, the residual gas must be scrubbed free of chlorine or liquefied in a second stage.

Figure 81 shows a two-stage liquefaction by the Uhde system, which operates at 0.3 – 0.4 MPa and – 20 °C in the first stage and – 60 °C in the second stage, with a yield of 99 % [221]. The refrigerant is difluoromonochloromethane. The gaseous refrigerant is compressed, liquefied by water cooling, and collected in a container. The liquid refrigerant is sprayed into the shell of the chlorine liquefier, where it evaporates, absorbing heat and cooling the chlorine, which flows from the liquefiers at – 15 °C (first liquefier) or – 55 °C (second liquefier) [222]. The residual gas from the first horizontal liquefier contains < 5 % hydrogen. It is fed into the second liquefier, which is at an angle of 60° and has a strong, low-volume construction. There the gas mixture passes through the explosive concentration limits. In case of an explosion, there is a comprehensive control system to ensure safety:

The explosion pressure is vented by means of a bursting disk to a residual gas absorber. Simultaneously the residual gas from the first stage is passed directly into this absorber. The chlorine gas to the second stage is shut off, and an inert gas purge is introduced. Finally, the liquid chlorine exit valve is closed to prevent back flow of the liquid chlorine into the second liquefier and from there into the absorber.

With normal-pressure (ca. 0.1 MPa) liquefaction and *low temperature* (< -40 °C), cryogenic storage of the liquid chlorine is possible. This process is advantageous when large quantities of chlorine must be liquefied as completely as possible. Attention must be paid to the increased solubility of other gases at low temperatures, especially carbon dioxide [207]. This carbon dioxide can be removed from the liquid chlorine by passage of hot chlorine gas [220].

An *absorption – desorption process* by Akzo is based on carbon tetrachloride [223]. It requires little energy and yields over 99.8 % of a pure liquid chlorine that is almost free of carbon dioxide. A similar process by Diamond Shamrock has been described [225].

11.6. Chlorine Recovery

Chlorine can be recovered from the tail gas from liquefaction with a chlorine recovery system.

Tail gas from liquefaction and chlorine from the plant evacuation system together with the snift compressor and stripper recycle streams are supplied to a snift compressor suction knock-out drum. The gas is compressed by the snift gas compressor to 7.0 kg/cm^2 with a discharge temperature of 85 °C.

The snift gas is then cooled by cooling water to 45 °C and then further cooled to -12.2 °C by Freon. Gas is sent to the absorber, whereas liquid is either returned to chlorine storage or is used for reflux at the stripper.

The off-gas enters the bottom of the chlorine absorber and passes upward through the two packed sections of the tower while cold carbon tetrachloride flows downward. All of the chlorine and the nitrogen trichloride is absorbed in the carbon tetrachloride while the noncondensable gases remain in the gasphase and are removed from the system.

The chlorine-rich carbon tetrachloride leaves the bottom of the chlorine absorber at ca. 10 °C and is forced by pressure difference to the chlorine stripper. Chlorine stripper feed enters the middle of the column and flows downward through two packed sections, releasing chlorine as it is heated. A thermosiphon reboiler is provided at the base of the stripper. By heating the liquid above 65 °C, the absorbed nitrogen trichloride decomposes to nitrogen and chlorine gas.

Chlorine boiled off in the stripper passes upward through a packed top section of the column where it is scrubbed and purified by liquid chlorine from the discharge knock-out drum. The stripper overhead stream, a mixture of chlorine and a small amount of inerts, is sent to the chlorine liquefaction system or recycled to the suction knock-out drum to maintain the stripper reflux [223], [224].

12. Chlorine Handling

Both the chlorine industry and governmental organizations are well aware of the risks of chlorine. In the United States and Canada, The Chlorine Institute [226] has established standards and recommendations for safe transport and handling of chlorine since 1924. In Europe, Euro Chlor, an association of major Western European chlorine manufacturers, publishes recommendations, codes, and memorandums for chlorine handling and transport concerning European conditions and regulations [227]. Both organizations distribute manuals and pamphlets worldwide. Surveys of existing national and international regulations for the handling and transport of hazardous chemicals are available [2], [34], [228], [230].

12.1. Storage Systems

Chlorine is liquified and stored at ambient or low temperature [231], [234]. In both cases the pressure in the storage system corresponds to the vapor pressure of liquefied chlorine at the temperature in the stock tank. *Pressure storage* is recommended for all usual customers [235], [236]. Euro Chlor recommends a maximum capacity of 300 – 400 t for individual tanks. For the large storage capacities required by producers, usually a *low-pressure storage* system, operating at a liquid chlorine temperature of ca. – 34 °C, is chosen. A low-pressure system needs a cooling or recompression system, and, for this reason, it is basically unsuitable for small chlorine consumers [237], [238].

A few major design aspects must be mentioned. Any risk of fire or explosion must be eliminated. All tanks having an external connection below the liquid level should be placed in a liquor-tight embankment (bund). In the event of leakage the liquid should be collected in a small area to reduce the rate of vaporization. The outer shell around a double-enveloped low-pressure storage tank can provide such a facility. To vent chlorine, there must be an absorption or liquefaction system. In the course of all operations, the design pressure should not be exceeded. The dimensions of branches and the amount of pipe work should be minimized. Bottom connections from storage tanks are not recommended for small chlorine users. Large branches should always be located in the gas space of a vessel. The pipework system should be provided with remotely operable valves to permit isolation in case of emergency.

Before being put into service, the whole storage system must be degreased, cleaned, and dried to achieve a dew point of – 40 °C in the purge gas at the outlet of the system. No substance that could react with the chlorine can be allowed to enter the storage system. The filling ratio in the tank should never exceed 95 % of the total volume of the vessel; for pressure storage tanks, this corresponds to 1.25 kg of liquid chlorine per liter of vessel capacity at 50 °C (Fig. 82).

Typical measuring and control equipment of a pressure storage tank is shown in Figure 83. The ISO codes for process measurement control functions and instrumen-

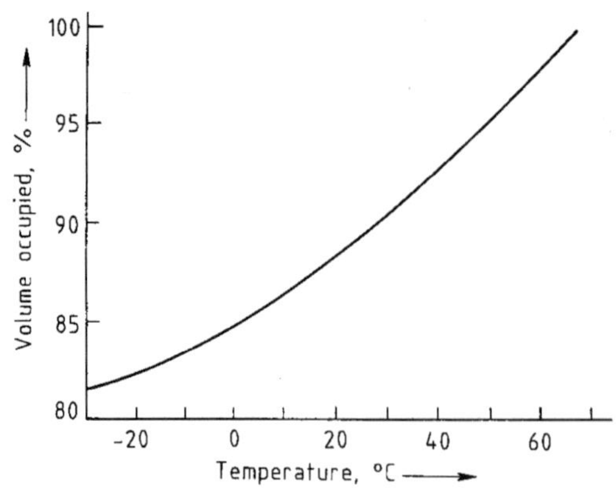

Figure 82. Proportion (%) of a one-liter vessel occupied by 1.25 kg of liquid chlorine as a function of temperature

tation are explained in Table 24 [239]. The measuring equipment of a low-pressure storage system needs supplementary devices, for example, a temperature indicator with an alarm and, in the case of a double-shell vessel, a device to determine the quality of the purging gas inside the double shell. The vessel and an external envelope should be protected against overpressure or underpressure. In low-pressure systems, the chlorine is removed by vertical submerged pumps, canned pumps below the vessel, or ejector pumps operating with a flow of liquid chlorine produced by external pumps.

Periodic inspection and retesting of the whole system, including a visual examination, a thickness test of the wall of the vessel and pipes, and an examination of the welds and the surfaces under any thermal insulation, is recommended. Hydraulic retesting is accompanied by risk of corrosion and is, therefore, not favored.

12.2. Transport

Within a chemical plant and over distances of several kilometers, chlorine can be transported by pipelines, either as gas or liquid [24], [240]. Every precaution should be taken to avoid any vaporization of chlorine in a liquid-phase system or any condensation in a gas-phase system. Wherever liquid chlorine could be trapped between two closed valves or wherever the system could be overpressurized by thermal expansion, an expansion chamber, a relief valve, or a rupture disk should be provided [241], [242].

Commercial chlorine is transported as a liquid, either in small containers (cylinders and drums) or in bulk (road and rail tankers, barges, and ISO containers). The design, construction, system of labeling, inspection, and commissioning are covered by national and international regulations [228]. *Cylinders* have a chlorine content up to 70 kg. A protective hood is provided to cover the valve during transport. The ton containers (drums) have a capacity of 500 – 3000 kg of chlorine. Drums are equipped with two valves near the center of one end and connected with internal eductor pipes.

Table 24. ISO codes and miscellaneous symbols for process measurement control functions and instrumentation

Codes	Function or Instrumentation
AA	analysis alarm
CW	cooling water
dPI	difference pressure indicating
FA	flowrate alarm
FI	flowrate indicating
FIA	flowrate indicating alarm
FICA	flowrate indicating controlling alarm
FRA	flowrate recording alarm
HZ	hand operated emergency acting
H	high
L	low
LA	level alarm
LIA	level indicating alarm
LIC	level indicating controlling
M	moisture analysis
PA	pressure alarm
PCZA	pressure controlling emergency acting alarm
PI	pressure indicating
PIA	pressure indicating alarm
PIAS	pressure indicating alarm switching
PIC	pressure indicating controlling
PRC	pressure recording controlling
PSA	pressure switching alarm
QRA	quality recording alarm
TA	temperature alarm
TC	temperature controlling
TI	temperature indicating
TIA	temperature indicating alarm
TIC	temperature indicating controlling
TRA	temperature recording alarm
WI	weight indicating
WIA	weight indicating alarm
⬭	measuring device
⋈⊸	remote control valve
– – –	control line

The capacity of *tank cars* (rail tankers) ranges from 15 to 90 t. Special angle valves are mounted on the manhole cover on top of the vessel. In Europe, pneumatic valves are normally used [231] – [233]. During loading and unloading, these valves can be closed rapidly and remotely in case of an accident. They have an internal safety plug, providing a tight seal against the passage of gas or liquid chlorine in the event of failure of the body of the valve.

In North America, the eductor pipe inside the vessel has an excess-flow valve at the top, immediately below the manhole cover. This valve closes the eductor pipe when the rate of liquid flow exceeds a set rate [2], [24]. North American tank cars have a spring-loaded safety relief valve, which protects the vessel against overpressure in case of external heat. The tanks have thermal insulation. In Europe thermal insulation and safety relief valves are not used or recommended.

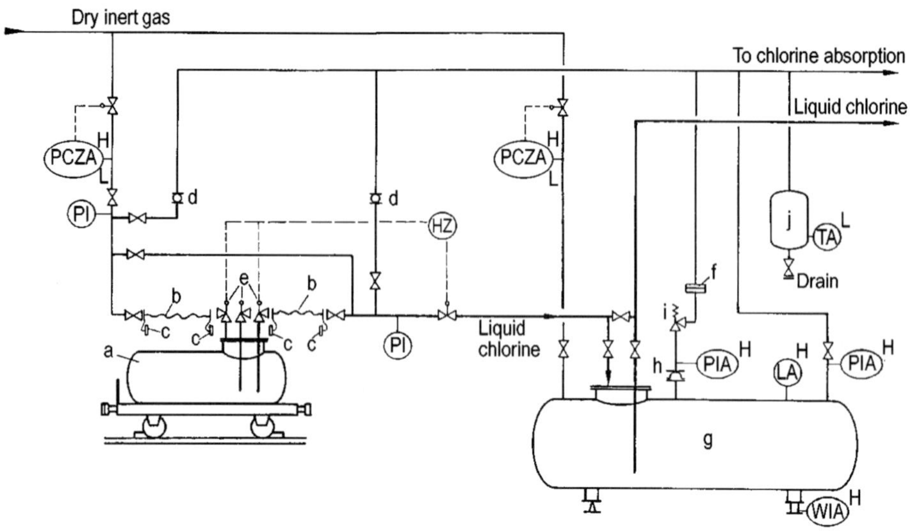

Figure 83. Discharge of liquid chlorine by padding to pressure storage
a) Liquid-chlorine rail tanker; b) Flexible connection; c) Plug; d) Viewing glass; e) Remote-control tank valves; f) Protective membrane; g) Storage vessel; h) Rupture disk; i) Relief safety valve; j) Buffer vessel for liquid chlorine

Road tankers and ISO containers have a chlorine capacity of 15 – 20 t. The design of and the equipment on chlorine pressure road tankers is similar to these of rail tankers. In North America, large amounts of chlorine are transported by *tank barges* [24]. These barges usually are of the open-hopper type with several cylindrical uninsulated pressure vessels. The total capacity of barges ranges from 600 to 1200 t. The chlorine is transported at low temperature.

Classification and Labeling. According to Directive EEC 67/548, Annex I, chlorine (Index no. 017-001-00-7) is classified as toxic and dangerous to the environment. The R and S phrases are:

R 23	toxic by inhalation
R 36/37/38	irritating to eyes, respiratory system and skin
R 50	very toxic to aquatic organisms
S 7/9	keep container tightly closed and in a well-ventilated place

For transportation, chlorine is in class 2, no. 2TC (toxic, corrosive) in ADR, RID, and ADNR in Europe, and in class 2.3 in the IMDG Code (p. 2116) and ICAO Code. All vessels must be labeled with the denomination 268/1017 Chlorine and with labels for dangerous goods: cylinders and drums with labels 2.3 (toxic gas)

ADR/RID: Nr.: 6.1

and 8 (corrosive)

railroad tankers with labels 2.3, 8, and 13 (shunt carefully); and for shipping, labels 2.3, 8, and MP (marine pollutant).

12.3. Chlorine Discharge Systems

All containers should be discharged in the same order as received. They must be placed where any external corrosion, risk of fire, explosion, or damage is avoided [24]. At normal room temperature, the discharge rate of chlorine gas from a single 70-kg cylinder is ca. 5 kg/h and the rate of a drum is ca. 50 kg/h. The flow of chlorine gas can be increased by a higher ambient temperature or by connecting two or more containers. A system of two or more containers must be carefully operated and controlled to avoid overfilling by transfer of chlorine from warm to cool containers. Direct heating of containers is not recommended [24]. The best way to determine the flow rate and container content is to observe the weight of the container [229]. A flexible tube is used to connect a mobile container with the fixed piping system. Any reverse suction from the consuming plant must be prevented by a barometric leg or other adequate precaution if the chlorination process runs at atmospheric pressure. Pressurized processes need a pressure controlling system with automatic isolation valves.

Uninsulated tanks have a maximum gaseous discharge rate of ca. 2 t/h. The chlorine gas can be used only for low-pressure chlorination processes and at low rates. This method increases the risk of concentrating nitrogen trichloride and other nonvolatile residues in the liquid phase within the tank. In all other circumstances, the liquid chlorine should be transferred into a fixed storage vessel and then vaporized in a special installation.

Liquid chlorine is discharged by putting the tank under pressure with dry inert gas or dry chlorine gas. The inert transfer gas must have a dew point below −40 °C at atmospheric pressure and must be clean and free of impurities such as dust or oil. Before closing the valves, the tanks must be vented to avoid the risk of high pressure in

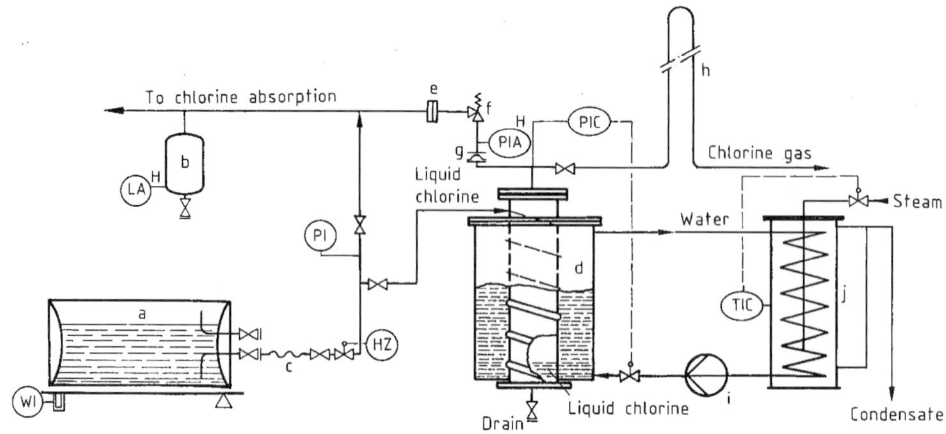

Figure 84. Liquid chlorine vaporizer
a) Liquid-chlorine drum; b) Buffer vessel; c) Flexible coil; d) Chlorine vaporizer; e) Protective membrane; f) Relief safety valve; g) Rupture disk; h) Barometric leg; i) Water pump; j) Water heater

the container on account of the additional partial pressure of the inert gas. The use of an inert gas requires the availability of a chlorine absorption or neutralization system. Discharge with pressurized chlorine gas requires a chlorine vaporizer or a special chlorine compressor. Articulated arms, flexible hoses, and steel coils are used for the flexible connections. Remote-control valves installed close to the ends of the flexible connections limit leakage in the event of a failure.

Recommendations on technical equipment, installation, taking into operation, checks, and handling are provided by The Chlorine Institute and Euro Chlor.

12.4. Chlorine Vaporization

When large amounts of chlorine gas are required or when the chlorination process needs pressurized gas, liquid chlorine must be vaporized and superheated to avoid liquefaction [243], [244]. It is advisable to operate the vaporizer at a sufficiently high temperature to accelerate the decomposition of nitrogen trichloride. As a source of heat, steam with a maximum allowed temperature of 120 °C is used when the vaporizing system is constructed of mild steel. Water above 60 °C is also suitable, as shown in Figure 84. Direct electrical heating is not appropriate because there is always a risk of overheating the steel.

Coil-in-bath vaporizers use a coiled tube or a spiral located in a vessel of hot water (Fig. 84). Generally, they are used for small throughputs; they are simple in design and construction. Double-envelope vaporizers have compact construction and are easy to operate and to maintain. Vertical tube vaporizers have a large surface area and allow a high flow rate. Kettle vaporizers are also constructed for large unit capacities [245].

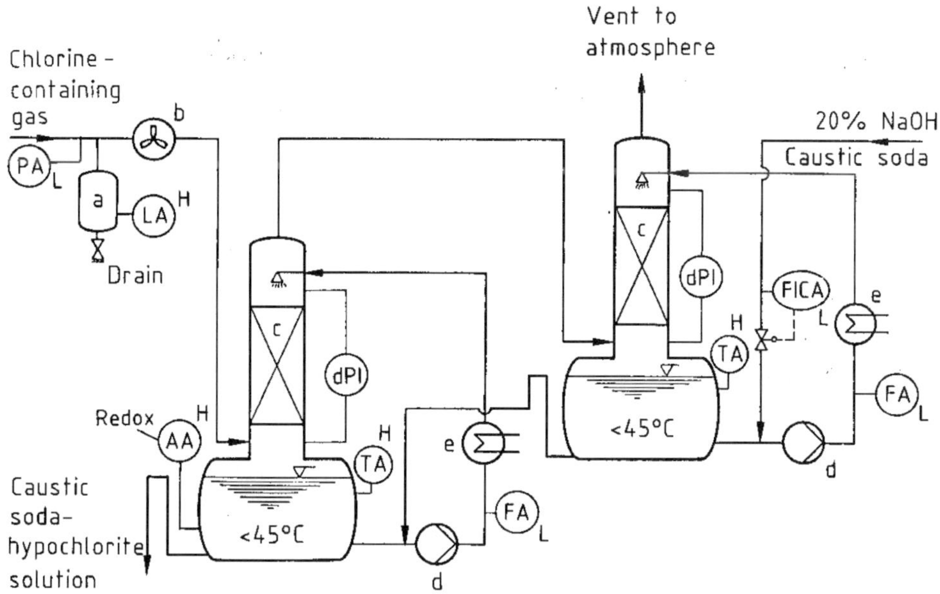

Figure 85. Absorption equipment for the treatment of gases containing chlorine
a) Buffer vessel; b) Vent fan; c) Packed tower; d) Circulating pump; e) Heat exchanger (cooler)

Every effort must be taken to avoid the reverse suction of water or organic materials into the vaporizer. The recommended water and nitrogen trichloride content of introduced liquid chlorine must not be exceeded. Vaporizers operating at low temperature or with a constant liquid level need to be purged to avoid dangerous concentration of nitrogen trichloride [246], [247].

12.5. Treatment of Gaseous Effluents

Gaseous effluents containing chlorine arise from various sources and must be treated in such a way as to obtain a tolerable concentration of chlorine when they are released into the air. The vent gas may contain other substances, such as hydrogen, organic compounds, CO_2, etc., which must be considered in design and operation of an effluent treatment installation [248], [249].

Operation of the collection system below atmospheric pressure facilitates the purging of chlorine vessels, pipes, etc. The risk of corrosion in dry chlorine installations by moisture from the treatment system must be excluded. The most commonly used and recommended reagent is caustic soda. The effluents are treated in an absorption system, such as packed absorption towers, venturi scrubbers, etc. An example of a flow sheet for a large plant is shown in Figure 85.

To avoid any formation of solid salts, the recommended concentration of caustic soda is < 22 wt %. The operating temperature should not exceed 55 °C; under normal conditions a temperature of ca. 45 °C is usual. A cooling system may be necessary. In large chlorine absorption units, the sodium hypochlorite solution that is produced can be used in other processes. Where this is not possible, several methods can be used to decompose the hypochlorite: controlled thermal decomposition, catalytic decomposition [250], acidification, for example, with sulfuric acid

$$NaCl + NaOCl + H_2SO_4 \longrightarrow Na_2SO_4 + Cl_2 + H_2O$$

12.6. Materials

The choice of material [251] depends on the design and operating conditions and must take into account all circumstances. A chlorine manufacturer should be consulted to confirm the suitability of a material. Any use of silicone materials in chlorine equipment should be avoided.

Dry Chlorine Gas (water < 40 ppm by weight). Carbon steel is the material most used for dry chlorine gas. It is protected by a thin layer of ferric chloride. For practical purposes the recommended temperature of these materials is ≤ 120 °C. High-surface areas, such as steel wool, or the presence of rust and organic substances increase the risk of ignition of steel.

The resistance of stainless steels to chlorine at high temperature increases with the content of nickel. For stainless steels containing less than 10 wt % nickel, the upper temperature limit is 150 °C. High-nickel alloys, such as Monel, Inconel, or Hasteloy C, are suitable up to 350 – 500 °C. Poor mechanical strength limits the use of nickel. Copper is used for flexible connections and coils, but it becomes brittle when stressed frequently.

Because titanium ignites spontaneously in dry chlorine, it must be avoided. Graphite, glass, and glazed porcelain are used where there is a risk of moisture in the dry chlorine gas, and poly(vinyl chloride) (PVC) or chlorinated PVC and polyester resins are suitable if the temperature limits of these materials are regarded.

Liquid Chlorine. Unalloyed carbon steel and cast steel are used with liquid chlorine. Low-temperature chlorine systems apply fine-grain steels with a limited tensile strength to guarantee good conditions for welding. To avoid erosion of the protective layer, practice is to limit the velocity of the liquid to less than 2 m/s. Organic materials — rubber lining, ebonites, polyethylene, polypropylene, PVC, chlorinated PVC, polyester resins, and silicone — are dangerous [252]. Zinc, tin, aluminum, and titanium are not acceptable. For certain equipment, copper, silver, lead, and tantalum are appropriate.

Wet Chlorine Gas. Wet chlorine gas rapidly attacks most common metallic materials with the exception of tantalum and titanium. To assure a protective oxide layer on the surface of the titanium, sufficient water must be present in the chlorine gas. If the system does not remain sufficiently wet, titanium ignites spontaneously [253].

Most organic materials are slowly attacked by wet chlorine gas. Rubber-lined iron is successfully used up to 100 °C. At low pressure and temperature the use of plastic materials like PVC, chlorinated PVC, and reinforced polyester resins is advantageous. Polytetrafluoroethylene (PTFE), poly(vinylidene fluoride) (PVDF), and fluorinated copolymers like tetrafluoroethylene – hexafluoropropylene (FEP) are resistant even at higher temperature. Ceramics have been progressively replaced by plastics. Impregnated graphite is suitable up to 80 °C; the impregnation should be resistant to wet chlorine.

Materials for Special Parts. After the ban of asbestos as material for gaskets, substitutes are recommended [254]. In wet chlorine gas, rubber or synthetic elastomers are acceptable. Even at temperatures up to 200 °C, PTFE is resistant against wet and dry chlorine gas and liquid chlorine.

Materials resistant because of protection by a chloride surface layer are not recommended for protective membranes, rupture disks, and bellows. Suitable materials are tantalum, Hasteloy C, PTFE, PVDF, Monel, and nickel.

12.7. Safety

In hazard and risk assessment studies, the design of chlorine installations and equipment and the operating and maintenance concepts are examined in detail to minimize risks [255]. However, there remains a certain risk, and all efforts must be taken to protect people and the environment in the case of a chlorine emergency. The penetrating odor and the yellow-green color of a cloud indicate chlorine in the air. If around-the-clock surveillance by operators is not possible, automatic leak detectors are available. Safety in handling chlorine depends largely on the education and training of employees. An emergency plan should be brought to the attention of the personnel involved. Computer-assisted systems can be used in certain circumstances [256]. Periodic exercises and safety drills should be carried out.

All people on a chlorine plant are advised to carry escape-type respirators. The use of filter masks is prohibited where there is a risk of a high concentration of chlorine. Anyone who enters an area with high chlorine concentrations should be equipped with self-contained breathing apparatus and full protective clothing suitable for dealing with liquid chlorine. Protective equipment, safety showers, eye-wash facilities, and emergency kits [24] must be quickly accessible.

A means of indicating the actual wind direction should be located near the chlorine installation.

Fixed or mobile water curtains can be used to divert the dispersion of a chlorine gas cloud [257]. However, the direct discharge of water into liquid chlorine and on the area of a chlorine leak must be avoided.

In most countries, chlorine manufacturers have organized groups of experts who are well versed and drilled in handling chlorine and can be called at any time in case of chlorine emergency.

The Chlorine Institute has released pamphlets and recommendations covering all aspects of safety, e.g., first aid [258], emergency response plans [259], protective equipment [260], prevention of injuries to personnel [261], prevention of chlorine releases [262], and estimating the area affected by a chlorine release [263]. Emergency kits have been developed for sealing leaks in chlorine containers, drums, and tank cars [264].

Euro Chlor offers a Chlorine Safety Manual [265], recommendations for emergency intervention [266], and for safe design, construction, operation of equipment [267].

Emergency plans have been established for accidents during transportation and use, e.g., CHLOREP [268] in North America and TUIS (Transport Unfall Informations System) in Germany.

13. Quality Specifications and Analytical Methods

13.1. Quality Specifications

Liquid chlorine of commercial quality must have a purity of at least 99.5 wt % [269]. The water content is < 0.005 wt %, and solid residues are < 0.02 wt %. The impurities are mainly CO_2 ($\leq$ 0.5 wt %), N_2, and O_2 (each 0.1 – 0.2 wt %). There are traces of chlorinated hydrocarbons (originating from rubberized or plastic piping) and inorganic salts such as ferric chloride. The chlorine may also contain small amounts of bromine or iodine, depending on the purity of the salt used in the electrolytic process.

13.2. Analytical Methods

Industrial liquid chlorine is mainly analyzed by the methods in ISO regulations. The liquid chlorine is evaporated at 20 °C, and this gas is then analyzed.

Sampling	ISO 1552 [270], [271]
Moisture	ISO 2121 [272], ASTM E410 [273]
	ISO 2202 [274]
Chlorine content	ISO 2120 [275], ASTM E412 [276]
Gaseous components	DIN 38 408, part 4 [277]
NCl_3	[278]
Mercury	[279], ASTM E506 [280]

The residue is weighed, and the organic constituents are taken up in acetone, hexane, or diethyl ether and determined by gas chromatography. The inorganic residue is analyzed. For quick analysis, liquid chlorine can be introduced directly onto a silica gel column of a gas chromatograph.

Chlorine Gas. The chlorine gas can be analyzed for chlorine content, gaseous impurities, hydrogen, organics, and moisture:

1) Chlorine content. One method for process monitoring and control of chlorine concentration is measurement of thermal conductivity.
2) Gaseous impurities. A known amount of chlorine gas is passed through a solution of potassium iodide or phenol to absorb the chlorine. The residual gases (O_2, N_2, H_2, CO, CO_2) are collected in a gas burette, measured, and analyzed by gas chromatography or with an Orsat apparatus.
3) Hydrogen. A known amount of air is added to the residual gas after removal of the chlorine to ensure excess oxygen, and the volume reduction is measured after the

hydrogen is consumed on a heated platinum coil. The hydrogen content can be continuously monitored by thermal conductivity measurement.

4) Organics. Organic components can be determined most conveniently by gas chromatography.
5) Moisture. A known amount of chlorine gas is passed through a drying tube filled with a weighed amount of phosphorus pentoxide. The moisture content is determined from the weight gain of the drying tube (ISO 2121). Continuous determination can be carried out, e.g., by absorption with phosphorus pentoxide and measuring the current required to electrolyze the absorbed water or by the electrical conductivity after absorption in sulfuric acid.

Detection of Chlorine. Chlorine can be recognized by smell or color. Small amounts can be detected by the blue coloration of starch – iodide paper, although other oxidizing agents can produce the same effect. Another method for chlorine detection depends on its ability to combine with mercury. If the unknown gas mixture is shaken with water and mercury, all the chlorine disappears and the remaining water has a neutral reaction. However, if the chlorine contains some hydrogen chloride, the water becomes acidic and reacts with silver nitrate solution to give a white precipitate (AgCl) that is soluble in aqueous ammonia. Leaks in pipes or equipment are detected by testing with the vapor from aqueous ammonia: a thick white cloud of chloride forms.

Quantitative Determination of Free Chlorine. The gas mixture can be shaken with a *potassium iodide solution,* and the liberated iodine can be then determined by titration. Chlorine in alkaline solution can be reduced to chloride by potassium or sodium *arsenite,* and the arsenite can be then oxidized to arsenate. The end point is detected by spot tests with starch – iodide paper. Excess arsenite is back-titrated with acidified potassium bromate solution. Small amounts of chlorine, e.g. in drinking water, can be determined by *photometric measurement* of the yellow color produced by the reaction with *o*-tolidine in hydrochloric acid solution [281].

To determine both *chlorine and carbon dioxide,* the chlorine is absorbed by a solution that contains known amounts of acid and potassium arsenite, and the chlorine is determined by back-titration of the arsenite. The carbon dioxide, which is not absorbed by this solution, is then absorbed by potassium hydroxide solution.

Detection Tubes. Commercial detection tubes (Drägerwerk, Lübeck; Auergesellschaft, Berlin) are available for measuring chlorine in air. They have various ranges: 0.2 – 3, 2 – 30, and 50 – 500 ppm. The Chlorometer (Zeiss-Ikon, Berlin) can determine the free chlorine content of water in a few minutes.

For the protection of the environment and control of working conditions, traces of chlorine as small as 0.01 – 10 ppm must be determined. There are many types of apparatus on the market for measuring workplace concentrations or emissions. They depend on physicochemical methods, such as conductometry, galvanometry, potentiometry, colorimetry, and UV spectroscopy [282].

14. Uses of Chlorine

The first industrial use of chlorine was to produce bleaching agents for textiles and paper and for cleaning and disinfecting. These were liquid bleaches (solutions of sodium, potassium, or calcium hypochlorite) or bleaching powder (chlorinated lime). Chlorine was then regarded merely as a useful chemical agent.

Since 1900s, chlorine has achieved constantly increasing importance as a raw material for synthetic organic chemistry. Chlorine is an essential component of a multitude of end products, which are used as materials of construction, solvents, pesticides, etc. In addition, it is contained in intermediates that are used to make chlorine-free end products. It is these areas of use that allow chlorine production to increase.

The percentage of world chlorine consumption of various product groups in 1997 was as follows [284]:

Vinyl chloride	33
Misc. organic products	6
Solvents	19
Pulp and paper	5
Water treatment	6
Others (inorganic products, etc)	31

The number of possible reactions of chlorine, and therefore the number of intermediates and end products, is remarkably large. Some important reactions are shown in Figure 86 along with the areas of application of the end products [285]. Figure 87 also shows some of the applications of chlorine.

The content of the following section is mainly based on the relevant articles in the 5th. and 6th. edition of Ullmann's Encyclopedia of Industrial Chemistry. A short description of the chemical reactions, some hints on the industrial practice and on the applications of the endproducts are given. Details may be obtained from the cited literature.

Considering the great variety of possible reactions in the chemistry of chlorine, only those applications are mentioned, which are important for industrial practice and/or are significant chlorine consumers.

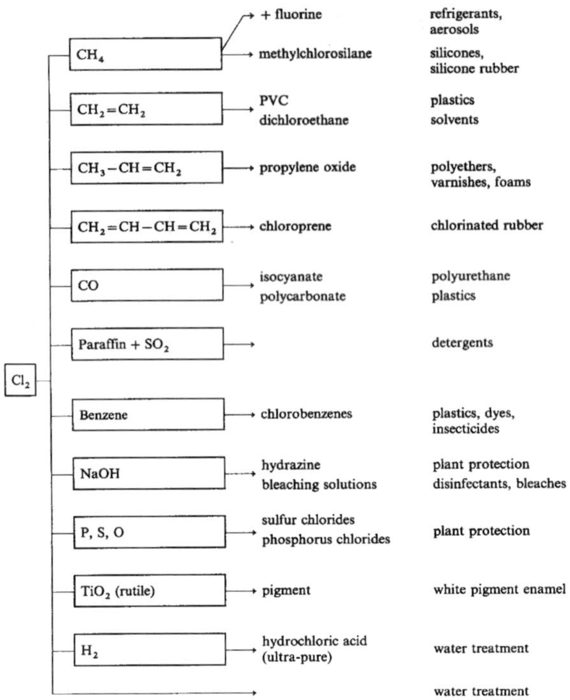

Figure 86. Important reactions of chlorine and the uses of the end products

14.1. Use of Elemental Chlorine

14.1.1. Water Disinfection

Safe drinking water has been recognised for centuries as a major determinant of health. Diseases arising from the ingestion of pathogens (bacteria, virus, protozoa) in contaminated water have the greatest health impact worldwide [320]. In 1991 the failure to disinfect drinking water had been a major contribution factor to the spread of a cholera epidemic from Peru to 16 other countries in Latin America, causing more than 20 000 fatalities. In 1990 diarrhoea was associated with 3.2 million death children under five years and additional one million deaths in older age groups [321]. About 80 % of all diseases and over one-third of death in developing countries are caused by the consumption of contaminated water and, on average, as much as one-tenth of each person`s productive time is sacrificed to water-related diseases.

Since the beginning of the twentieth century chlorine gas or liquid chlorine is used as disinfectant. Today about 98 % of West Europe's drinking water is purified by chlorination.

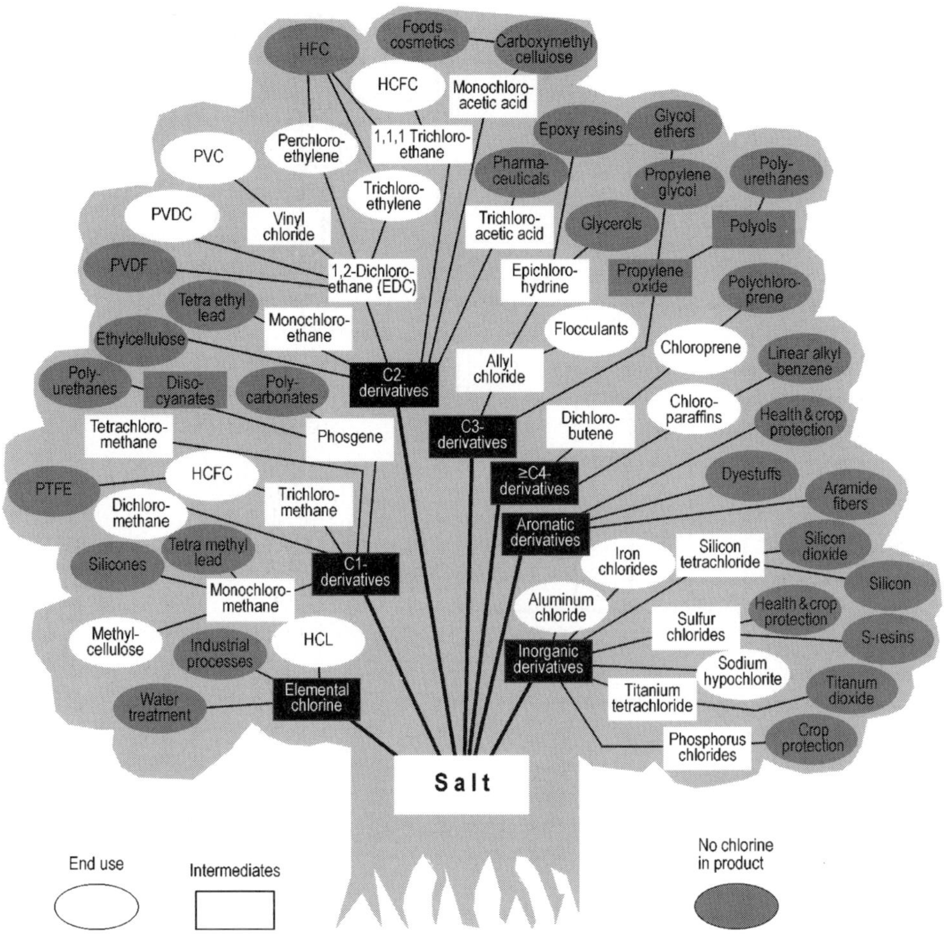

Figure 87. The chlorine tree. (Courtesy of Euro Chlor)

When chlorine is added to water, a mixture of hypochlorous and hydrochloric acid is formed

$$Cl_2 + H_2O \longleftrightarrow HOCl + H^+ + Cl^-$$

The equilibrium depends on the pH level: at pH < 4, the equilibrium is displaced to the right and little Cl_2 exists in solution, the chlorine exists predominantly as HOCl. Between pH 6.0 and 8.5, the HOCl dissociates

$$HOCl \longleftrightarrow H^+ + OCl^-$$

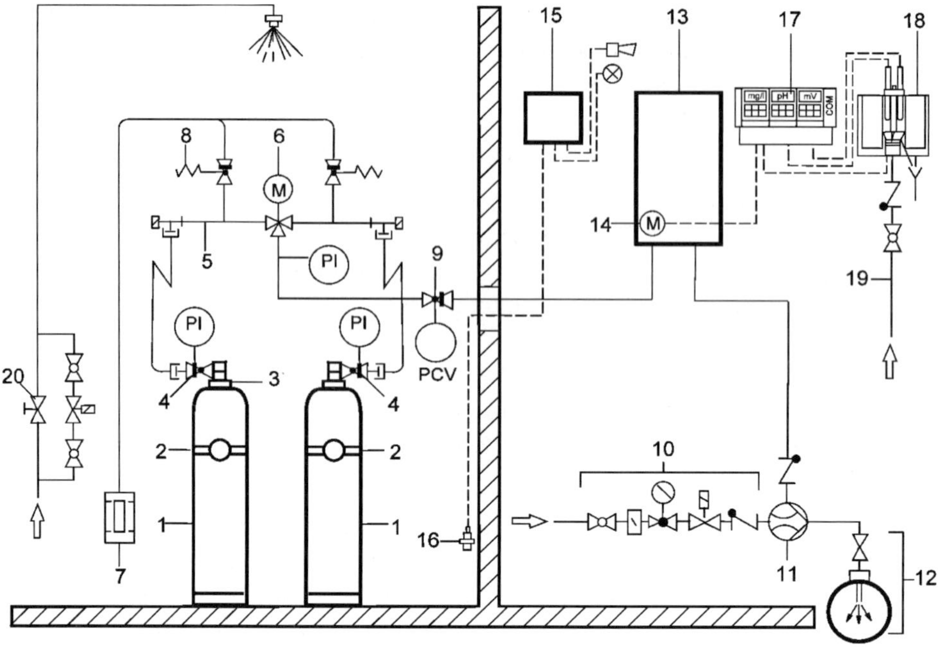

Figure 88. Application of chlorine into drinking water.
1. Chlorine cylinders 2. Fixing of the cylinders 3. Cylinder valve 4. Vacuum control- and non-return-valve 5. Vacuum pipe for chlorine 6. Automatic switching valve 7. Absorption filter 8. Safety degassing valve 9. Vacuum safety valve 10. Chlorine water device 11. Chlorine injector 12. Chlorine water/drinking water mixing device
13. Vacuum chlorine dosage apparatus 14. Electric control motor 15. Chlorine gas alarm 16. Chlorine gas sensor 17. Measurement and control modules for pH-value, redox-value and chlorine concentration 18. Flow through device with electrodes 19. Water for analysis, continuous flow taken after a reaction time of 30 minutes 20. Water sprinkler

Above pH 7.8 hypochlorite ions (OCl-) predominate, and they exist almost exclusively at pH > 9.

Chlorine existing in water as hypochlorous acid and hypochlorite ions (or both) is defined as *free available chlorine.*

Hypochlorous acid is a considerably more efficient disinfectant than hypochlorite ions, thus efficient disinfection is favored by lower pH.

A description of the application of chlorine into drinking water is given in Fig. 88. It shows a modern installation with the chlorine cylinders in a separate room, all chlorine gas pipings between cylinders and injector working under vacuum, with all measuring-, controlling- and safety devices for the dosage (System USF Wallace & Tiernan).

Chlorine and its derivative water treatment products, sodium hypochlorite and chlorine dioxide, rapidly destroy bacteria and other micro-organisms. In addition to its disinfecting properties, chlorine prevents the growth of algae and slime in pipes and storage tanks. As a further advantage it has a residual disinfection effect, and it is the only technique which can assure disinfection right up to the tap. The residual ability to destroy and inhibit the activity of pathogenic agents is a specific characteristic of chlorine. Therefore the dosage of chlorine into the raw water is choosen high enough to maintain — after the reaction of the chlorine with the water and all ingredients — a residual chlorine concentration of 0.2 to 0.3 mg/L in the distribution system to prevent microbial recontamination.

Chlorine reacts with chemical substances in the water:

- with the formation of chlorite and chlorate
- with ammonium ions it forms chloramines
- with organic nitrogen and sulphur compounds it forms organic chloramines and other chloro -addition and -substitution products
- metal ions (iron, manganese) are oxidized, nitrites to nitrates, sulfur to sulfate
- the presence of organic substances, e.g. humic acids, can lead to the formation of undesired by-products, e.g. trihalomethanes (chloroform, bromodichlormethane, dibromochlormethane, bromoform, also trichloracetic acid), some of these are considered to be carcinogens [322].

Alternative water disinfection techniques use chlorine dioxide, ultraviolet radiation, ozone and ultrafiltration. Normally these techniques form less hazardous by-products, but the technologies are more complex and the total costs are higher. Little is known about the nature and possible toxicity of the other disinfectants and their by-products and they do not have the ability of the residual disinfection effect, so the treated water may be protected from recontamination by a small dose of chlorine. The quality of the raw water and the concentration and the nature of the contaminants may justify the decision for an alternative disinfection technique.

With modern chlorine dosage installations as described above, it is possible to use the chlorine in such a moderate manner, that the limit for trihalomethanes of 10 micrograms per liter is not exceeded, but the safe disinfection is ensured [323].

Taking into consideration the minimal concentrations of hazardous by-products resulting from chlorination, the World Health Organization (WHO) comes to the conclusion, that the risks to health from chlorine disinfection by-products are extremely small in comparison to the risks associated with inadequate disinfection [324].

Because of hygienic requirements swimming pool waters and certain municipal and industrial waste waters are also chlorinated.

Swimming pool water must be disinfected similar to drinking water. The fast reaction of the chlorine or the hypochlorous acid kills most of the germs within a few seconds — and the residual disinfection effect is especially useful in this case. Difficulties in breathing or the typical "chlorine smell" is caused by chlorine disinfection when it destroys organic contaminants like sun oil, sweat, tear glands, saliva, hairs and even urine from swimmers. In order to prevent difficulties, the chlorine industry has developed recommendations for users [325], [326]. Initiatives try to help pool users to improve the pool hygiene by educating users and improving pool design.

The chlorination of drinking water and of public swimming pool water normally is done by the application of chlorine gas from drums or cylinders. An alternative is the electrochlorination, which consists of producing a bleach solution on site by the electrolysis of a salt solution [327]. The handling of chlorine is done by skilled personnel, which is trained for this application. This handling could be a danger for small, private owned swimming pools, therefore the use of sodium hypochlorite solution or tablets of dry calcium hypochlorite is recommended for this purpose.

Waste water chlorination is even older than drinking water chlorination. It makes use of the toxicological, oxidative and coagulant properties of chlorine. Besides the main target, the disinfection, chlorine is used to deodorize the waste water, e.g in presence of hydrogen sulfides, to decolorize, to retard putrefaction, to reduce the biochemical oxygen demand (BOD), and to ease the filtration by improved coagulation.

In industrial waste waters, chlorine can react with ingredients like phenols to form hazarduous derivates. In such cases other disinfectants should be used. For each of these applications, a risk - benefit study has to be performed.

Whenever an overchlorination has taken place, the surplus of free chlorine can be reduced by addition of sodium thiosulfate pentahydrate (Antichlor). 0.88 grams of sodium thiosulfate per m^3 of water reduces the chlorine content by 1 mg/L.

The direct use of elemental chlorine for sterilizing water has declined in some areas but not in others. For example, in Germany the percentage is $< 0.1\,\%$ of the production, but in the United States it is ca. 5 % (1995) [313].

14.1.2. Pulp and Paper

The pulp and paper industry produces paper, cellulose and its derivates from wood.

Wood and other lignocelluloses such as straw consist mainly of cellulose, hemicelluloses, and lignin, which acts as a binder. In a "cooking" process according to the sulfite or sulfate (Kraft) process, the lignin and the semicelluloses are mostly dissolved out of the fiber matrix. The fibrous material obtained thereafter is called pulp. For most uses, the residual lignin (ca 2 – 5 %) and colored or color producing substances must be eliminated from the cellulose in a subsequent bleaching treatment [328].

In conventional pulp bleaching, chlorine, hypochlorite, and chlorine dioxide are used with or without intermediate alkaline extraction steps. The first step in conventional lignin-dissolving bleaching is the chlorination. Chlorine water is normally used for this chlorination. It converts the residual lignin to products, that are soluble in water and/or alkali. Chlorine and hypochlorite react primarily with the benzene or phenol rings of lignin, in which substitution and oxidation reactions take place. In a separate extraction step, the unwanted soluble substances are removed with water and/or diluted sodium hydroxide solution.

The effluents from the bleaching process contain chlorinated organic compounds and cannot be disposed of by combustion with the spent cooking liquor. It pollutes the waste water stream with its oxygen demand and its toxic and genotoxic effects, it may even contain polychlorinated dibenzo-furans and dibenzo-p-dioxins in small concentrations. The stability of the chloro-organic compounds increases its biological and environmental persistence and this increases the tendency to bioconcentration in organisms. The concentrations of the resulting organochloric compounds sometimes presented an unacceptable risk to the environment and the food chain.

Chlorine is an very effective, selective and inexpensive bleaching agent, but the brightening of the pulp is not sufficient. Therefore, as a third step, the brightening was effected by chlorine dioxide, which is extremely effective and selective in brightening, but is also expensive.

The production of the chlorine dioxide is carried out at the pulp mill by the reduction of sodium chlorate in dilute aqoeous solutions

$$NaClO_3 + 2\,HCl \rightarrow ClO_2 + 1/2\,Cl_2 + NaCl + H_2O$$

The bleaching effect of chlorine dioxide is mainly based on the oxidation of lignin. The formation of chlorinated organic compounds is sharply reduced compared to chlorine.

In response to pollution linked to the use of chlorine, paper mills are converting their bleach process to

- the substitution of chlorine dioxide for chlorine in the prebleaching process: ECF-Bleach = Elemental Chlorine Free Bleach
- the developing of chlorine free bleaching processes by using alkali/oxygen, peroxide or ozone: TCF-Bleach = Total Chlorine Free Bleach.

Both of the alternatives have the following advantages:

The amount of absorbable (AOX) and extractable organic halogen (EOX) decrease to lower levels (e.g. from Swedish paper mills decreases in AOX from 2.8 kg/tonne of pulp to < 0.2 kg are reported), the amount of chlorophenols decreases, the remaining chlorophenols are chlorinated to a lower extent, less chlorinated acetic acids and less chloroform are formed. The chlorine compounds are biodegradable and do not persist in the environment.

The TCF-pulp has no advantages over the ECF-pulp. Because all wood contains organic chlorine derivatives it is technically impossible to produce chlorine free paper from wood pulp.

Future developments of the paper mills aim at closed circle versions of the pulp bleaching processes, so completely avoiding discharges to the environment.

Uses of pulp: 95 % of all pulps are "paper pulps", used for paper and board production, from the rest, the "dissolving pulps", chemicals are made like viscose (cord rayon, modal fibers, cellulose guts, cellulose films, acetyl cellulose) or cellulose derivatives (cellulose ethers, cellulose nitrates, methyl cellulose, carboxy cellulose). The requirements concerning quality differ considerably from one application to the other.

World paper production was 44 million tonnes in 1951, it rose to 228 million tonnes in 1988, and it is still rising. Main producing areas are North America (United States 30 % share, Canada 7 %), Europe (West 26 %, East 7 %), and Japan (11 %). The per capita consumption in1988 was highest in the United States (310 kg), followed by Sweden

(247 kg), Japan and West European countries (between 200 and 210 kg). In Latin America it was 25 kg, in Africa 5 kg and in some developing countries < 1 kg.

In 1990 about 10 % of the world`s chlorine production (ca. 3.4 million tonnes) was used for pulp and paper [329]. As a consequence of the change-over to chlorine free bleaching processes, the use of elemental chlorine was drastically reduced to 2.95 million tonnes in 1994, and is anticipated to be 1.9 million tonnes in 2000) [319]. In 1983 in the United States 13 % of the chlorine production was used for pulp and paper, 1994: 9 %, 2000: 6 %, 2010: 0 % (estimated). In areas like Canada or Western Europe, the reduction was even more drastic: 1998, in Western Europe 18 000 tonnes of chlorine went into this production, accounting for only 0.2 % of the production [330].

14.2. Inorganic Nonmetal Chlorides

14.2.1. Phosphorchlorides [331]

Phosphorus trichloride is the most important phosphorus-halogen. The world production capacity was estimated (1988) at > 300 000 t. It is formed by the combustion of phosphorus in a dry stream of chlorine

$$P_4 + 6\ Cl_2 \longrightarrow 4\ PCl_3 \qquad H = -1276\ kJ/mol$$

In a continuous process, molten white phosphorus and gaseous chlorine react in previously produced phosphorus trichloride.

Uses : PCl_3 is an important intermediate in the production of phosphate ester insecticides. It is used as a chlorinating agent and as a catalyst. Phosphorus pentachloride, phosphoryl chloride, thiophosphoryl chloride and phosphonic acids are made from PCl_3.

Phosphorus pentachloride, PCl_5, [*10026-13-8*] can be prepared by the reaction of excess chlorine with PCl_3

$$PCl_3 + 6\ Cl_2 \longrightarrow PCl_5 \qquad H = -125\ kJ/mol$$

It is a chlorinating agent in organic chemistry. In the pharmaceutical industry it is used in the manufacture of penicillin and cephalosporin. Further it is used to produce acid chlorides and as a catalyst for cyclisation and condensation reactions.

Phosphorus trichloride oxide (phosphoryl chloride), $POCl_3$ [*10025-87-3*] is obtained by oxidising PCl_3 with air or oxygen.

$$PCl_3 + 1/2\ O_2 \longrightarrow POCl_3 \qquad H = -2975\ kJ/mol$$

It is used to produce phosphate esters by reaction with alcohols and phenols, acid chlorides, dyes as triphenyl methane.

Thiophosphoryl chloride, $PSCl_3$ [*3892-91-0*], is prepared by direct reaction of excess sulfur with PCl_3 at ca 180 °C. It is used to introduce the P-S-group in organic compounds to give thiophosphate esters. The inorganic phosphorus chlorine compounds are starting materials for numerous organic phosphorus compounds, e.g. phosphines, halophosphines, phosphonium salts, phosphonous and phosphinous and phosphonic acids and their derivatives and finally esters of phosphoric acids [332].

The world production in 1988 was ca 170 000 t PCl_3, 4000 t PCl_5, and 80 000 t $POCl_3$.

14.2.2. Sulfur Chlorides [333]

Disulfur dichloride, S_2Cl_2, [*10025-67-9*], the most stable sulfur chloride, is formed by passing gaseous chlorine into molten sulfur at temperatures of 220 to 260 °C, the gaseous S_2Cl_2 leaving the reactor is then condensed

$$2\,S + Cl_2 \longrightarrow S_2Cl_2$$

S_2Cl_2 is used in the production of cutting oils and additives for high-pressure lubricating oils, for vulcanizing agents, for rubber compounds, for organosulfur compounds, pharmaceuticals and crop protection agents. The production in 1992 was ca. 15 000 t.

Sulfur dichloride, SCl_2, is made by passing gaseous chlorine into S_2Cl_2 at low temperature. Antioxidants, organosulfur compounds and rubber chemicals are made using it. Production 1992 ca. 8000 t.

Thionyl chloride, $SOCl_2$ [*7719-09-7*], is one of the most important chlorinating agents in organic chemistry. It is produced by treating sulfur dioxide or sulfur trioxide with SCl_2 at 150 – 250 °C in the presence of activated carbon.

$$SO_3 + SCl_2 \longrightarrow SOCl_2 + SO_2$$

$$SO_2 + SCl_2 + Cl_2 \longrightarrow 2\,SOCl_2$$

Thionyl chloride is used for the production of acid chlorides from carboxylic acids, of alkyl chlorides from alcohols, for ester production and for chloro-methylation. Areas of use are crop protection, pharmaceuticals, dyes, paper and textile auxiliaries, and plastics. Production 1992 ca. 45 000 t.

Sulfuryl chloride, SO_2Cl_2, [*7791-25-5*] is formed by reacting SO_2 and chlorine over activated charcoal with cooling

$$SO_2 + Cl_2 \longrightarrow SO_2Cl_2.$$

It is used in the production of chlorophenols or in side-chain chlorinations of aromatics, of sulfonic acids and the sulfochlorination of polyethylene. In lithium batteries it acts as liquid cathode.

Not directly related to chlorine, is the chlorosulfonic acid, produced by the reaction of hydrogen chloride with sulfur trioxide

$$HCl + SO_3 \longrightarrow HSO_3Cl$$

which is an important intermediate for the production of detergents, pharmaceuticals, dyes, crop protection, ion exchange resins, plastics and others. The annual production is ca 250 000 t.

14.2.3. Nitrogen – Chlorine Compounds

Chloroamines [334]. Inorganic chloroamines are prepared in pH-controlled reactions by the action of hypochlorous acid or chlorine on ammonium salts, e.g.

$$NH_3 + xHOCl \longrightarrow NH_{3-x}Cl + xH_2O$$

They are widely used as disinfectants and bleaching agents for paper.

N-Chloroisocyanuric acids, chloroisocyanurates, are prepared by continous reaction of chlorine with isocyanuric acid in aqueous sodium hydroxide at 0 – 15 °C. Their oxidizing and bioactive properties derive from the hypochlorous acid that is slowly released from them in water. They are used as disinfectants for swimming pools, cleaning and sterilizing of bathrooms, laundry bleach and nonshrinking treatment of wool.

Other organic chloroamines serve as bactericides, disinfectants, chlorinating agents in organic synthesis, and bleaching agents.

Cyanuric chloride is prepared by chlorinating hydrocyanic acid at 20 – 40 °C in aqueous medium, and after drying, trimerizing to cyanuric acid.The most important cyanuric acid derivates are the aminotriazines, which are used as pesticides, especially as herbicides [335].

Hydrazine and its derivates are used as blowing agents for plastic foams, boiler water treatment, polymerization initiators, pesticides, pharmaceuticals, and dyes.

In the Raschig process for the production of hydrazine, ammonia is oxidized at 5 °C with sodium hypochlorite to give monochloroamine in a first step

$$NaClO + NH_3 \longrightarrow NH_2Cl + NaOH$$

In a second step, the reaction mixture is mixed with a large molar excess of ammonia and the slow reaction to hydrazine is carried out at 130 – 150 °C and 3.0 MPa

$$NH_2Cl + NH_3 + NaOH \longrightarrow N_2H_4 + NaCl + H_2O$$

From the diluted reaction liquor the ammonia is evaporated, and the salt removed in a forced circulation salting-out evaporator. The distillate is then concentrated to 100 % hydrazine hydrate [336].

Nitrogen trichloride, NCl_3, is explosive even in small concentrations. It can be formed, whenever nitrogen containing compounds, e.g. amines, ammonium, ammonia, are chlorinated. Therefore, the chlorinating conditions must be chosen in a way, which excludes this risk. The handling of this problem in chlor-alkali electrolysis plants is discussed in detail in Chapter 11.

Nitrosyl chloride, NOCl, is an important chemical for nitrosification, oximisation, diazotation in the oil refining industry and in the production of dyes (diazotation with NOCl instead of nitrite), it serves as a catalyst in isopron polymerisation and as a non-aqueous solvent. It is formed from nitric oxide and chlorine

$$2\ NO + Cl_2 \longrightarrow 2\ NOCl.$$

The nitrosyl process (See chapter 10.2.2) is one way to recover chlorine from hydrogen chloride by oxidizing with oxygen or air.

14.2.4. Hydrogen Chloride, HCl, and Hydrochloric Acid

Hydrogen chloride and its aqueous solution, hydrochloric acid [337], are two of the most important basic industrial chemicals. Very pure HCl gas is formed by the direct synthesis of the gaseous elements

$$H_2 + Cl_2 \longrightarrow 2\ HCl$$

in a combustion chamber. The hot product, T > 2000 °C, is cooled, purified if necessary, and either liquified and filled in bottles or tankers, or supplied to the user on site by pipes. Very pure hydrochloric acid is usually produced by absorbing the synthesis HCl gas in purified water.

The reaction of concentrated sulfuric acid with common salt or potassium chloride produces sodium or potassium sulfate, and as a by-product, HCl.

$$H_2SO_4 + 2\ NaCl \longrightarrow Na_2SO_4 + 2\ HCl.$$

Huge quantities of HCl are obtained from the chlorination of organics by substitution

$$R\text{-}H + Cl_2 \longrightarrow R\text{-}Cl + HCl,$$

from the cracking of 1,2 dichloroethane to give vinyl chloride and HCl

$$C_2H_4Cl_2 \longrightarrow C_2H_3Cl + HCl$$

or from the production of non-chlorine containing chemicals from chlorine containing compounds, e.g. polyurethanes and polycarbonates with the use of phosgene.

Incineration of chlorinated organic waste is another source of HCl.

The amount of hydrogen chloride produced as a by-product exceeds by far the demand of the market for hydrochloric acid. Therefore most producers in the chlorine consuming industry try to recirculate the HCl into the production process, as far as it is economically feasible, e.g. by the oxychlorination process in the vinyl chloride production

$$C_2H_4 + 2\ HCl + 1/2\ O_2 \longrightarrow C_2H_4Cl_2 + H_2O$$

producing 1,2 dichloroethane from ethane and hydrogen chloride by catalytic oxidation with air, by electrolysing the hydrochloric acid to get back hydrogen and chlorine

$$2\ HCl \longrightarrow H_2 + Cl_2$$

by neutralizing it with alkali to produce salts, e.g. $CaCl_2$, NaCl, KCl or finally to sell the hydrochloric acid.

The management of the hydrochloric acid streams plays a decisive role for the equilibrium of the chlorine balance of an economic region. It is estimated, that in Europe one third of the total chlorine demand is covered by HCl [338], see Chapter 14.8.

14.2.5. Oxygen Chlorine Compounds [339]

From numerous chlorine oxides, only chlorine dioxide, ClO_2, is fairly stable under certain conditions. The chlorine oxygen acids (hypochlorous acid, chlorous acid, chloric acid and perchloric acid) are formed by reaction of the corresponding chlorine oxides with water.

Dichlorine oxide, Cl_2O, and chlorine dioxide, ClO_2, all oxygen acids and their salts, particularly those of sodium and potassium, are used industrially. Their significance is based predominantly on their oxidizing power. The strongest oxidants are those with the lowest oxidation state of the chlorine atom. For 200 years, the textile and paper industry has been a main consumer of ClO_2 and hypochlorite.

Hypochlorous acid is produced by the reaction of chlorine and water

$$Cl_2 + H_2O \longrightarrow HOCl + HCl$$

Its use as water disinfectant is discribed in chapter 14.1 (Use of elemental chlorine).

Solid hypochlorites are stable up to 80 °C, their stability depending on the water content, which is $< 1\,\%$ for bleaching powder, $< 0.3\,\%$ for tropical bleach. They decompose by reaction with water or by heating to 180 °C.

Bleaching powder, $Ca(OCl)_2$, is produced by passing chlorine over hydrated lime

$$5\ Ca(OH)_2 + 2\ Cl_2 \longrightarrow Ca(OCl)_2 \cdot 2\ Ca(OH)_2 + CaCl_2 \cdot Ca(OH)_2 \cdot H_2O$$

The dried product has a content of 35 – 37 % of available chlorine. It is used as an bleaching agent in single-stage bleaching and for disinfection purposes.

Hypochlorite solutions are prepared by reacting chlorine gas with alkali hydroxide solutions, e.g. 20 % caustic soda solutions. The available chlorine concentrations are 170 – 220 g/L (14 – 15 %).

$$Cl_2 + 2\ NaOH \longrightarrow NaOCl + NaCl + H_2O$$

or electrochemically by on-site decomposition of brine or sea water (av. chlorine < 10 g/L). A great number of companies are offering hypochlorite cells for producing different lots of available chlorine and chlorite concentrations, and for other applications.

Uses of diluted hypochlorite solutions are in the pulp and paper and the textile industry for bleaching, water disinfection, biofouling control in desalinating projects and power generation sites, disinfection of seawater for secundary oil recovery, food production.

Chlorine dioxide, ClO_2, [*10049-04-4*] is an extremely unstable gas, decomposing readily into chlorine and oxygen even on mild heating. It is explosive as a gas or liquid at high concentration. However, it can be handled easily when it is diluted with air to < 15 %. Because of the explosive risk, ClO_2 is manufactured on-site immediately before use. The industrial production is based on the reduction of chlorate with sulfur dioxide, hydrochloric acid, or methanol as reducing agents

$$2\ NaClO_3 + 4\ HCl \longrightarrow 2\ ClO_2 + 2\ NaCl + Cl_2 + 2\ H_2O$$

$$2\ NaClO_3 + H_2SO_4 + SO_2 \longrightarrow 2\ ClO_2 + Cl_2 + 2\ NaSO_4 + 2\ H_2O$$

$$2\ NaClO_3 + CH_3OH + H_2SO_4 \longrightarrow 2\ ClO_2 + HCHO + Na_2SO_4 + 2\ H_2O$$

Small-scale consumers can produce ClO_2 by passing chlorine gas through sodium chlorite solution

$$2\ NaClO_2 + Cl_2 \longrightarrow 2\ ClO_2 + 2\ NaCl$$

or by reaction of hydrochloric acid with sodium chlorite

$$5\ NaClO_2 + 4\ HCl \longrightarrow 4\ ClO_2 + 2\ H_2O + 5\ NaCl$$

Uses: The importance of ClO_2 is steadily rising, replacing elemental chlorine as a bleaching agent in the pulp and paper and the textile industry and as a specific disinfectant.

Sodium chlorite, $NaClO_2$, [*7758-19-2*] is industrially produced by treating ClO_2 with caustic soda and the simultaneous reduction of the formed chlorate to chlorite, for example using hydrogen peroxide

$$2\ ClO_2 + 2\ NaOH + H_2O_2 \longrightarrow 2\ NaClO_2 + 2\ H_2O + O_2$$ [340]

The product, a 33 wt % solution, is used as a bleaching agent in textile industry.

Sodium chlorate, $NaClO_3$, is industrially produced by electrosynthesis. The world production capacity is more than 10 million tonnes per year, and so it is one of the most important electrochemical production processes.

An aqueous sodium chloride solution is electrolysed in a cell without diaphragm at 80 – 90 °C, pH 6.1 – 6.4. Hypochlorite forms as an intermediate, that is further oxidized to chlorate either within the cell (ca 20 %),

$$3\ ClO^- + 1.5\ H_2O \longrightarrow ClO_3^- + 3\ H^+ + 2\ CL^- + 0.75\ Cl^- + 3\ e^-$$

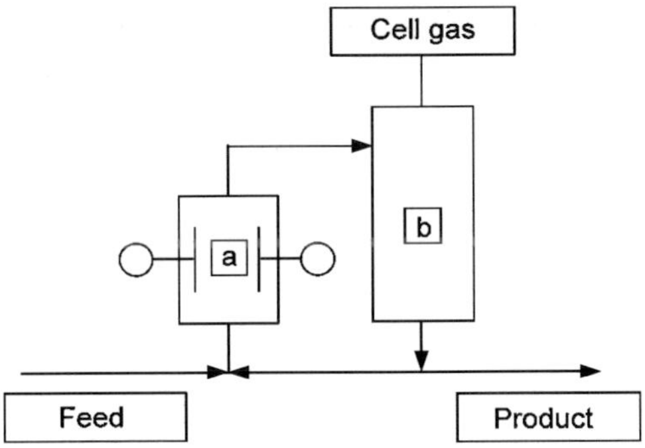

Figure 89. Electrochemical production of sodium chlorate
a) Electrochemical reactor; b) Chemical reactor

or by autooxidation in a separate reactor

$$3\ HClO \longrightarrow ClO_3^- + 2\ Cl^- + 3H^+$$

The concept of the separate electrochemical reactor (a) and the chemical reactor (b) is shown in Fig. 89.

The reactor product contains 550 – 750 g/L sodium chlorate and 90 – 100 g/L sodium chloride at 85 – 95 °C. Solid sodium chlorate is obtained from the solution by crystallisation after flash cooling or evaporation.

Potassium chlorate is generated from sodium chlorate and potassium chloride

$$NaClO_3 + KCl \longrightarrow KClO_3 + NaCl$$

Uses: The growth of the production of sodium chlorate is mainly due o the change-over from elemental chlorine to chlorate-based chlorine dioxide in the pulp and paper industry. Other uses are the production of potassium chlorate, sodium perchlorate, ammonium perchlorate for solid propellants, as oxidizing agent in uranium refining, as additive to agricultural products and to dyes, in textile and fur dyeing, metal etching. Potassium chlorate is used for the manufacture of matches, in pyrotechnics, explosives, cosmetics and in the pharmaceutical industry.

Perchlorates are commercially produced by the anodic oxidation of sodium chlorate in aqueous solution

$$ClO_3^- + H_2O\text{-}2\ e^- \longrightarrow ClO_4^- + 2\ H^+$$

Other perchlorates are gained by converting the sodium perchlorate

$$NaClO_4 + KCl \longrightarrow KClO_4 + NaCl$$

$$NaClO_4 + NH_4Cl \longrightarrow NH_4ClO_4 + NaCl$$

Perchloric acid is used in the analytical chemistry and as an acetylation catalyst for cellulose and glucose. Ammonium perchlorate serves as an oxidizing component in solid rocket propellants. Lithium perchlorate is used in dry batteries, potassium perchlorate in pyrotechnics, magnesium perchlorate as a drying agent.

14.3. Metal Chlorides

Metal chlorides are produced either as endproducts like $FeCl_3$, $AlCl_3$, or as intermediates for non-chlorine containing products, like titanium metal. The chlorinated intermediates allow the separation and purification of raw materials to give end products with a very high purity.

14.3.1. Titanium Chlorides [341]

Titanium dioxide is one of the most important pigments in the color producing industry.

Titanium metal is used as material for the construction of aircraft, submarines, chemical apparatus, e.g. anodes for the chlor-alkali electrolysis, and in medicine for protheses and surgery instruments, for watches and so on.

Until 1960, the digestion of titanium ores was performed by the sulfate process, using sulfuric acid. Severe environmental problems with acidic waste waters led to the development of the chloride process, which is increasingly replacing the sulfate process. 1999 one-third of the TiO_2-production is using the chloride process. The titanium industry is a growing consumer for chlorine.

The raw materials, rutile sands, synthetic rutiles, and titanium slags contain high concentrations of titanium dioxide. The chlorination process is carried out in a continuous fluidized-bed process at temperatures of 750-1000°C in the presence of carbon (e.g. calcined petroleum coke)

$$TiO_2 + 2\,Cl_2 + 2\,C \longrightarrow TiCl_4 + 2\,CO$$

The gaseous crude titanium(IV)-chloride, boiling point 136 °C, is separated by fractionated condensation from other metal chlorides ($SiCl_4$, $SnCl_4$, $VOCl_2$, $FeCl_3$, $AlCl_3$) and purified by vacuum distillation. The metal chlorides are neutralized with lime and deposited as hydroxides [342].

The pure $TiCl_4$ is burned with air or oxygen to give pure TiO_2 and chlorine gas

$$TiCl_4 + O_2 \longrightarrow TiO_2 + 2\,Cl_2$$

The chlorine gas is recirculated into the chlorination process. The TiO_2 is further processed to give pigments. In 1995, 2 % of the chlorine production of the United States about 200 000 tonnes, were used for the production of TiO_2.

Pure titanium metal is produced from $TiCl_4$ by metallothermic processes using sodium metal (Hunter process) or magnesium metal (Kroll process) as reductive agents at temperatures of 800 – 850 °C

$$TiCl_2 + 2\ Mg \longrightarrow Ti + 2\ MgCl_2$$

The titanium metal is obtained as crude titanium sponge, which is purified from magnesium by vacuum distillation to give pure titanium sponge. The magnesium metal and the chlorine gas is recovered by molten salt electrolysis.

Main producing countries for titanium sponge are the United States, Japan, Russia, Kazakhstan, China and India. The world capacity in 1994 was > 113 000 tonnes per year.

$TiCl_4$ is used also for the production of artificial smoke, of Ziegler – Natta – catalysts, and it is a starting material for titanium acid esters and for organic titanium compounds.

Titanium trichloride, $TiCl_3$, is manufactured from $TiCl_4$ by reduction with a surplus of hydrogen in red hot glowing tubes

$$TiCl_4 + H_2 \longrightarrow TiCl_3 + HCl$$

It is used as a reduction medium for the reduction of nitrogen-oxygen compounds, as a component in Friedel – Craft – reactions in the rubber industry, as a bleaching agent for azo dyes in the textile industry and as an indicator in chemical analytics.

Titanium dichloride, $TiCl_2$, is produced from titanium trichloride by thermal decomposition, or by the reduction of titanium tetrachloride with sodium amalgam.

14.3.2. Zirconium Chloride [343]

Zirconium metal finds application as construction material in nuclear power plants and in the chemical industry, and as a component of special steel alloys.

The manufacture of zirconium metal is similar to that of titanium: The raw materials zircon ($ZrSiO_4$) and baddeleyite (ZrO_2-containing ore) are processed by chlorination with coke as a reducing agent

$$ZrSiO_4 + 4\ Cl_2 + 4\ C \longrightarrow ZrCl_4 + SiCl_4 + 4\ CO$$

$$ZrO_2 + 2\ Cl_2 + C \longrightarrow ZrCl_4 + CO_2$$

at temperatures of 600-800°C in a fluidized bed reactor.

The $ZrCl_4$ is then purified by sublimation.

$ZrCl_4$ is mainly used for the preparation of Zr-metal sponge via reduction with alkali metal or magnesium metal.

14.3.3. Aluminum Chloride [344]

Aluminum chloride, $AlCl_3$, is prepared by the reaction of dry chlorine gas with liquid aluminum at 750 – 800 °C

$$2\ Al + 3\ Cl_2 \longrightarrow 2\ AlCl_3$$

or by the chlorination of aluminum oxide in presence of carbon.

The uses of aluminum chloride are manifold.

Water-free $AlCl_3$ acts as a strong Lewis acid in the organic chemistry, especially in Friedel-Crafts-reactions for the alcylation of aromatics, e.g. in the production of ethyl benzene. It is a catalyst for the production of ethyl chloride and is a precursor material for the production of dyes, detergents, resins, aluminum borohydride, lithium aluminum hydride, and of phosphorus and sulfur compounds.

Aqueous solutions of $AlCl_3$ are used as a flocculant in waste water treatment, as catalyst in the textile and paper industry, as a disinfectant and in wood protection.

The production of metallic aluminum by electrochemical decomposition of $AlCl_3$ (ALCOA-process) has not yet reached industrial scale.

14.3.4. Iron Chlorides [345]

Iron(II)-chloride, $FeCl_2$, is prepared by dissolving iron powder in diluted hydrochloric acid

$$Fe(s) + 2\ HCl \longrightarrow FeCl_2 + H_2$$

Water-free $FeCl_2$ is obtained by the reaction of dry hydrogen chloride gas with red hot iron powder. It is used as a reducing agent in the production of colours.

Iron(III)-chloride, $FeCl_3$, in water-free form is produced by the chlorination of hot (red heat) iron scrap

$$2\ Fe + 3\ Cl_2 \longrightarrow 2\ FeCl_3$$

It is used as an oxidizing agent in textile printing, as a coagulant and flocculant in water or waste water treatment, for graphitizing of coke, as etching agent for metals, in the preparation of dyes, and as catalyst in Friedel-Crafts-reactions.

14.3.5. Other Metal Chlorides

Metal chlorides with only a small-scale production are boron trichloride, antimony chlorides, tantalum pentachloride, tungsten chlorides, vanadium chlorides. Their properties, manufacture and uses are discribed in [346].

14.4. Silicon

The element silicon, Si, plays a decisive role in modern human life, and chlorine chemistry plays a decisive role in silicon chemistry, though chlorine is not a constituent in the applications of silicon and of most silicon compounds. Modern electronics is almost exclusively based on silicon devices, both in low-power and in high-power electronics. Silicon products have captured market share in many applications because of their superior performance [347].

The raw material for the production of silicon is mostly quartz sand, silicon dioxide, which is reduced by a carbothermical process

$$SiO_2 + 2\,C \longrightarrow Si + 2\,CO$$

in electrical furnaces at temperatures of 1800 – 2000 °C. The product is metallurgical grade silicon with a purity of ca. 98 %.

Elemental silicon for use in integrated circuits requires a purity of > 99.9999 % Si. This is achieved by converting the raw silicon into a chlorosilanes, SiH_2Cl_2, $SiHCl_3$, and $SiCl_4$. The raw silicon is milled to a sand or powder, fed into a fluidized bed reactor, where it is fluidized by a stream of hydrogen chloride at temperatures of ca 650 °C.

$$Si + 3\,HCl \longrightarrow SiHCl_3 + H_2$$

$$Si + 4\,HCl \longrightarrow SiCl_4 + 2\,H_2$$

The gaseous product, e.g. trichlorosilane, is filtered free from dust, condensed, and distilled in a multistage distillation free from low-boiling and high-boiling impurities (Fig. 90).

Semiconductor-grade silicon is produced by converting the silicon compounds to elemental silicon by chemical vapor deposition (CVD), employing the strongly endothermic reaction

$$4\,SiHCl_3 + 2\,H_2 \longrightarrow 3\,Si + SiCl_4 + 8\,HCl,$$

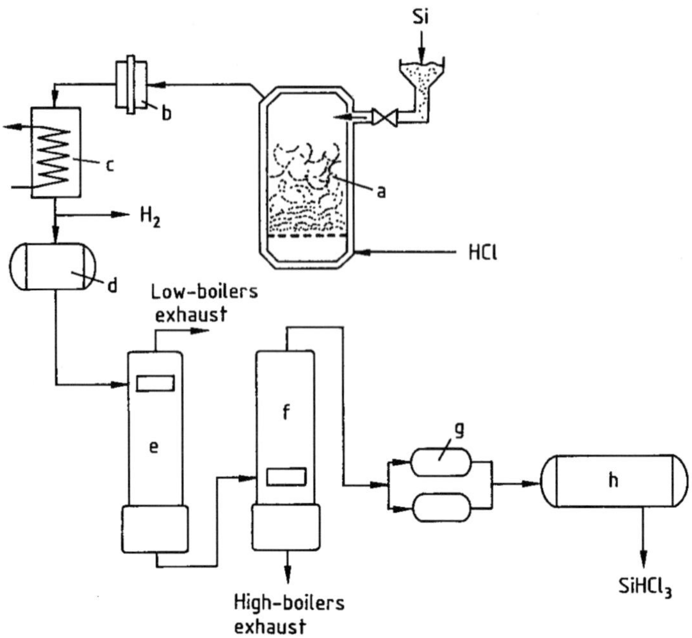

Figure 90. Flow chart of the preparation and refining of trichlorosilane
a) Fluidized-bed reactor; b) Dust filter; c) Condenser; d) Tanks; e) Distillation of low-boiling impurities; f) Distillation of high boilers; g) Tanks; h) Storage tanks

the back-reaction from above. The CVD takes place in an electrically heated, so-called "bell-jar reactor" at temperatures of ca. 1230 °C (see Fig. 91). The CVD includes a further purification step, because only highest purity silicon is deposited. Silicon rods are obtained with a length of >2 m and a diameter of <0.25 m.

The by-product $SiCl_4$ is converted back into trichlorosilane

$$SiCl_4 + H_2 \longrightarrow SiHCl_3 + HCl$$

Remaining silicon tetrachloride is used for the production of highly dispersed silicon dioxide or of synthetic silica glass. The evolving hydrogen chloride gas is recycled into the production of trichlorosilane.

Single crystal growth is achieved by the Czochralsky method, or the flat-zone method. The single crystals are sewn into wafers and further processed to give e.g. integrated circuits.

Photovoltaics, e.g. solar cells, are using monocrystalline, multicrystalline or amorphous silicon. The relevant production processes are developed to a high degree of sophistication.

Silicones are materials with a remarkable wide spectrum of applications. They are synthetic polymers, in which silicon atoms and oxygen atoms are arranged in the form

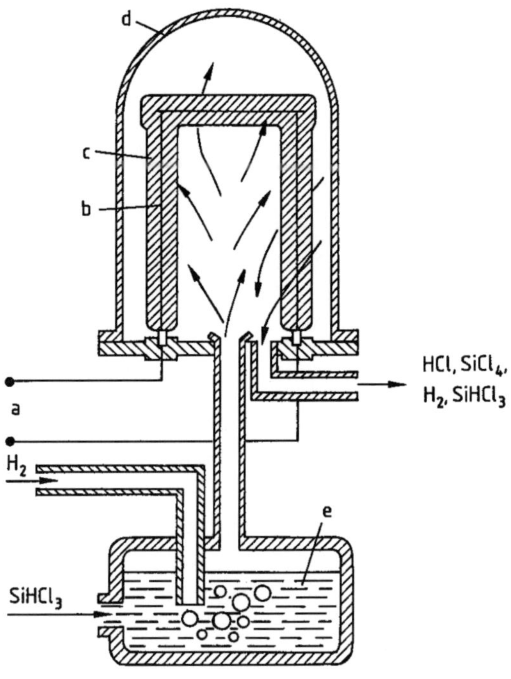

Figure 91. Chemical vapor deposition of silicon a) Electrical current; b) Starting silicon slim-rod; c) CVD polycrystalline silicon rod (1400 K); d) Reactor (silica, metal); e) Saturator

of chains or of networks. The free valences of the silicon atoms are saturated with organic hydrocarbons, mostly with methyl groups, more seldom with ethyl-, propyl-, or phenyl-groups.

The preparation of silicones is started from dimethyldichlorosilane, which is produced from silicon powder and methyl chloride by the Rochow-synthesis

$$Si + 2\,CH_3Cl \longrightarrow (CH_3)_2SiCl_2$$

In a fluidized bed reactor, copper serves as a catalyst, the temperature is about 300 °C.

Most silicones are polymers of dimethyl siloxane, which is produced by the hydrolysis of dimethyldichlorosilane

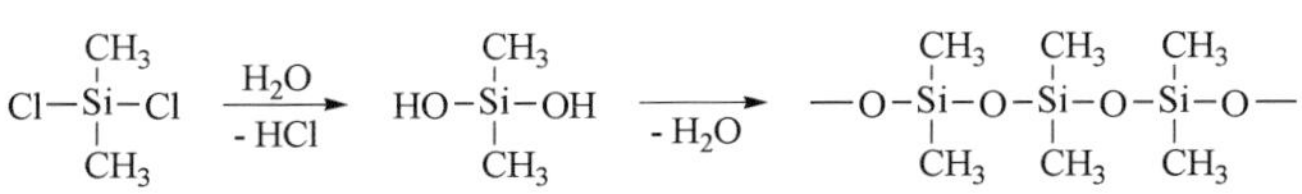

Other production processes make use of methanolysis instead of hydrolysis

$$(CH_3)_2SiCl_2 + 2\,CH_3OH \longrightarrow \text{-(-}(CH_3)_2SiO\text{-)-} + 2\,CH_3Cl + 2\,H_2O$$

The chlorine is obtained back from the hydrolysis process as hydrogen chloride, which is in turn reacted with methanol to give methyl chloride. In a principially closed loop, the methyl chloride is fed back into the Rochow synthesis.

Uses: Silicones are classified as fluids, elastomers or resins [348].

The fluids are chemically inert, resistant to attack by heat and oxidation, have low surface tension and impart water repellency when applied as coatings. They are used in cosmetics (shampoos, antiperspirants, cremes), pharmacological formulations and a wide range of industrial applications such as transformer fluids, polishes, waxes, antifoam agents, processing aids.

Silicone elastomers are used mainly as adhesives and sealants where high performance is required. They are used to obtain durable, tight seals in metal-to-glass joints in buildings, for example, and for other difficult-to-seal joints in the construction and industrial sectors. For automotive applications they are used as gaskets, seals, O-rings, and for protecting electrical components.

Silicone resins are used in applications requiring high thermal or chemical resistance, specific electrical properties, or water repellency, as coatings or as components of paints that withstand harsh conditions such as exposure to high temperature or corrosion in marine environments.

The great variety of properties is obtained by the use of different substituting groups in the siloxane synthesis, variation of the molecular weight, the extent of copolymerization with other polymers (alkyd resins, polyesters), and the addition of additives, pigments, fillers.

In 1990, ca. 800 000 tonnes of silicones were produced. The production and the number of applications is still rising.

14.5. Phosgene

Phosgene, $COCl_2$, is a characteristic example for a chlorine-containing precursor substance for the production of chlorine-free endproducts. These endproducts, especially the polyurethanes and the polycarbonates, have a great variety of applications in daily life. The significance of phosgene as a consumer of chlorine rises steadily. 1989 the world production was ca 2.7 million tonnes, the growth rate in the 90s was 4 %/year.

Commercially, phosgene is produced by passing carbon monoxide from a steam reforming plant and chlorine gas over activated carbon

$$CO + Cl_2 \longrightarrow COCl_2$$

The process is strongly exothermic, $\Delta H = -107.6$ kJ/mol, so the produced gas is cooled and either processed immediately or liquified by refrigeration (b.p. 7.56 °C at 101.3 kPa) and stored [349]. Noncondensed phosgene is absorbed directly with solvent to form phosgene solution. Nonabsorbable gases (CO, inert gases, phosgene) are fed to the waste-gas treatment (Fig. 92).

Phosgene is highly toxic. The necessary safety precautions include hermetically sealed equipment containing a small inventory, immediate use after production, auto-

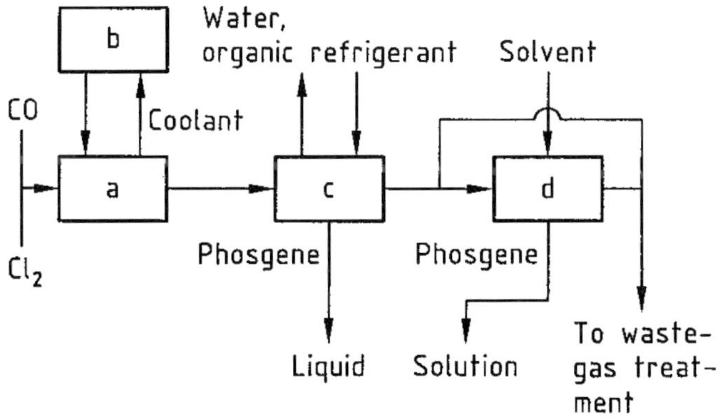

Figure 92. Simplified flow sheet for the production of phosgene
a) Reaction; b) Recooling of the coolants; c) Phosgene liquefaction; d) Phosgene absorption

matic shut down of the plant in emergency case, leak detectors and curtains of steam and ammonia around the plant to absorb escaping gases.

The commercially most important reaction of phosgene, covering 85% of its use, is with primary amines to form carbamyl chlorides, from which isocanates are formed by elimination of hydrogen chloride at elevated temperatures

$$R\text{-}NH_2 + COCl_2 \longrightarrow R\text{-}NH\text{-}COCl + HCl$$

$$R\text{-}NH\text{-}COCl \longrightarrow R\text{-}NCO + HCl$$

The two reactive chlorine atoms at the opposite ends of the phosgene molecule determine the use of phosgene for addition reactions, polymerization reactions and for chain-enlargements. Other uses of phosgene are reactions with secondary amines to give imidoyl chlorides, with tertiary amines to form cationic complexes, with nitriles to produce heterocycles and with metal oxides to produce metal chlorides.

In 1998, the worldwide annual production of polyurethanes was about 7 million tonnes, that of polycarbonates was about 1.1 million tonnes [350]. Therefore, these applications are here discussed in detail.

Polyurethanes [351]. The production starts from toluene, which is converted by HNO_3 in concentrated sulfuric acid to give a mixture of 2,4- and 2,6-dinitrotoluene (DNT). Catalytic hydrogenation with hydrogen gas gives toluene diamines (TDA). The reaction of TDA with phosgene forms toluene diisocyanate (TDI) (see Fig. 93).

The production of methylenediphenyl diisocyanate (MDI) starts from aniline, which is condensed with formaldehyde to a mixture of isomeric methylenediphenyldiamines (MDA). The reaction of MDA with phosgene is similar to that of TDI.

JOHN RYLANDS
UNIVERSITY
LIBRARY OF
MANCHESTER

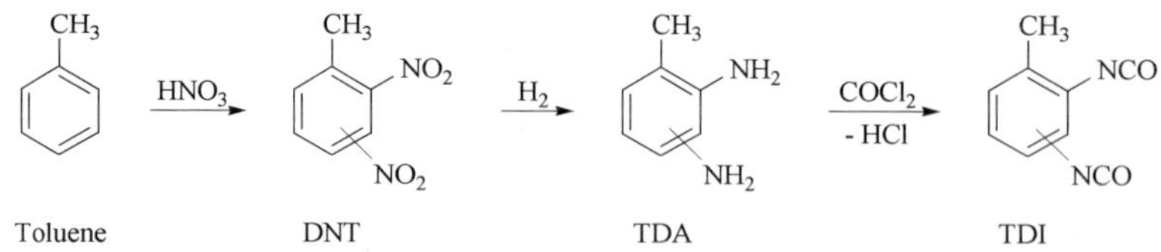

Figure 93. Preparation of toluene diisocyanate (TDI)

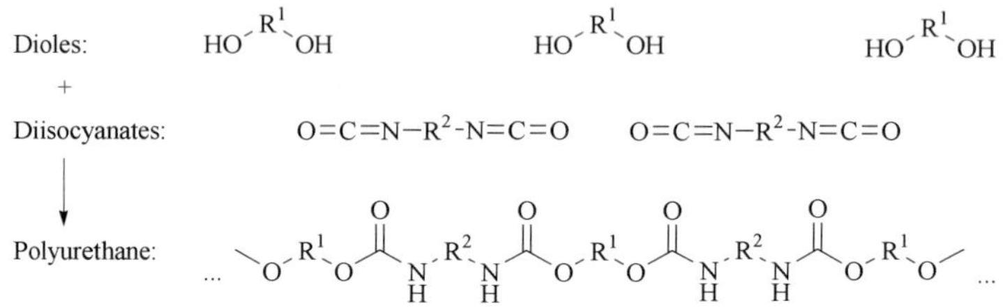

Figure 94. Preparation of polyurethane — basic reaction is polyaddition

The byproduct hydrogen chloride is absorbed in water to give hydrochloric acid, which can be electrochemically decomposed to give chlorine gas — to be recycled into the phosgene production — and hydrogen gas, which is used in the hydrogenation process.

The polyaddition of diisocyanates with di-oles or poly-oles (e.g. polyalkylenglykol ether) formes polyurethanes, the properties of which can be varied in a very wide range by the combination of the reaction components (see Fig. 94).

The high toxicity of phosgene has resulted in intensive work on the development of phosgene-free processes for diisocyanate manufacture. The development has been successful for only a few aliphatic diisocyanates, but ca. 95 % of all isocyanates produced in an industrial scale are aromatic.

Polyurethanes are supplied as solutions, aqueous dispersions, powders, microcapsules, hydrogels, flexible and rigid foams, plates. Uses are construction parts in civil engineering and car construction, flat roof gulleys, dome lights, window frames, instrument housing, vending machines and cable distributor boxes, thermal insulating material in houses, cars and refrigerators, as leather imitates, and they are components of adhesives, coatings and lacquers. As an example, Figure 95 shows possible uses of PURs in a passenger car.

Polycarbonates (PC) are thermoplasts. Their great commercial success is due to their unique combination of properties: extreme toughness, outstanding transparency, compatibility with other polymers, high heat distortion resistance, and high electric resistivity [352].

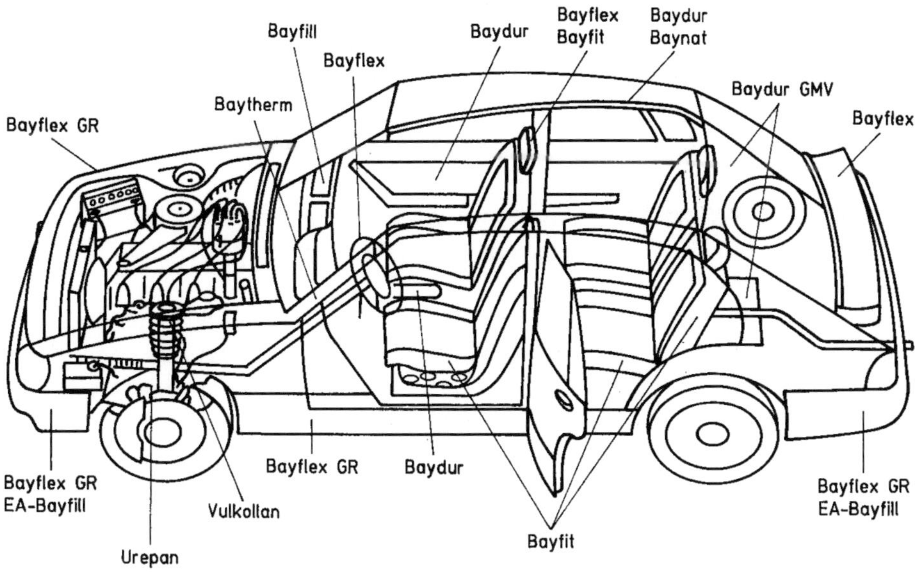

Figure 95. Use of PURs in the passenger car
Baydur = rigid integral foam; Bayfill = semiflexible filling foam (behind sheets); Bayfit = flexible molded foam; Bayflex = semiflexible integral foam; Baynat = hot-moldable rigid foam; Baytherm = rigid foam; EA = impact-energy-absorbing; GMV = glass-mat reinforced; Urepan = polyurethane rubber; Vulkollan = PUR hot-coating elastomer

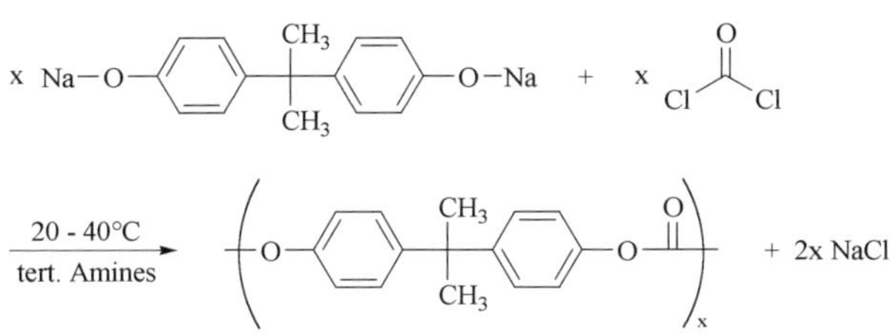

Figure 96. Preparation of biphenol A polycarbonate — overall reaction

The economically most important PC is the bisphenol A polycarbonate (2,2-bis(4-hydroxy-phenol)propane polycarbonate).

In the industrial production of this PC, interfacial polycondensation is used. The bisphenol A is first dissolved in the aqueous phase as sodium salt, and the phosgene in the organic phase, which is not miscible with water, e.g. dichloromethane. The reaction occurs at the interface of the two phases to produce oligomers, which enter the organic phase. The hydrolysis product NaCl enters the aqueous phase. The addition of catalysts (tertiary amines) accelerates the polycondensation process. The chlorine leaves the process as sodium chloride, see Fig. 96.

The properties of PC's can be adjusted to the requirements of customers by copolymerization with other polymers and/or addition of additives, e.g. mould-release agents, flame retardants, light stabilizers, reinforcements, heat stabilizers, colours.

Uses of polycarbonates: The electric sector consumes ca. 44% of the produced PC. Applications are housings for telephones, distribution equipment, lamp sockets, safety switches,...

The building and construction industry use the transparency and toughness of PC for making window panes and roofing for railway stations, houses, greenhouses, windows for telephone booths.

In the automobile sector (8%), light covers, reflectors, dashbords are made.

Compact discs (CD`s) and ROM-systems are made from PC, also office equipment (casings, covers, keybords), films and sheets (cheque cards), and among others, foams, packaging, baby bottles, camera lenses, and optical fibers.

14.6. Chlorinated Aliphatic Hydrocarbons

From the great number of chlorinated aliphates, only significant consumers of chlorine are described in this chapter.

14.6.1. Chloromethanes [353]

The industrial preparation of chloromethane derivatives is based to a wide extent on the treatment of methane and/or monochloromethane with chlorine, whereby the chlorination products are obtained as a mixture of the individual stages of chlorination: CH_3Cl, CH_2Cl_2, and CCl_4.

In Germany, ca. 20% of the produced chlorine was used in the chlorination of methane in 1992.

Thermal chlorination is preferred, but photochemical or catalytic methods are also employed. The thermal chlorination is a radical chain reaction, initiated by chlorine atoms at temperatures of 350 – 550 °C

initiation step	$Cl_2 \longrightarrow 2\,Cl'$
chain propagation	$CH_4 + Cl' \longrightarrow CH_3' + HCl$
	$CH_3' + Cl_2 \longrightarrow CH_3Cl + Cl'$
chain termination	$2\,Cl' + M \longrightarrow Cl_2 + M$

(where M = walls, impurities, oxygen)

The product distribution in methane chlorination is shown in Fig. 97 as a function of the ratio chlorine : methane. This distribution can be influenced by working with a high methane to chlorine ratio, by admixing inert gases (nitrogen), recycled hydrogen chloride or monochloromethane into the feed gas, and by proper temperature control.

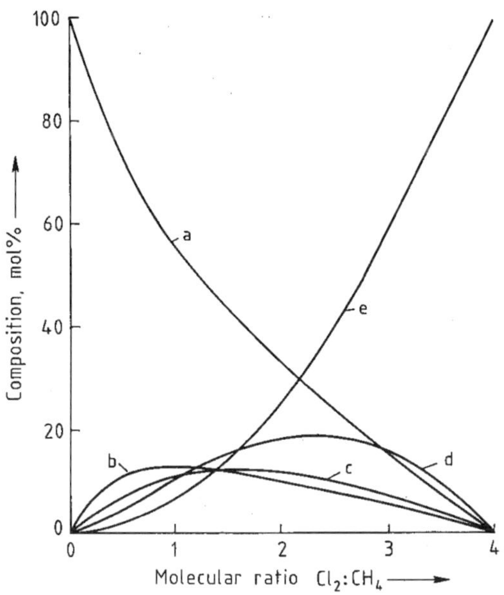

Figure 97. Product distribution in methane chlorination, ideal mixing reactor
a) Methane; b) Monochloromethane;
c) Dichloromethane; d) Trichloromethane;
e) Tetrachloromethane

In this way the explosion range of the methane-chlorine mixture can be avoided, as well as decomposition reactions of the chloromethanes, which take place when a critical temperature of ca. 550 – 700 °C is exceeded.

14.6.1.1. Monochloromethane, Methyl Chloride, CH_3Cl

The importance of the isolation of monochloromethane from the thermal chlorination reaction by distillation is declining. A specific way to produce CH_3Cl is the reaction of methanol and hydrogen chloride

$$CH_3OH + HCl \longrightarrow CH_3Cl + H_2O$$

The hydrochlorination process is carried out catalytically in the gas phase at 0.3 – 0.6 MPa and temperatures of 280 – 350 °C with activated aluminum as catalyst. It leads to a single target product, and has the advantage of consuming HCl.

The natural occurrence of methyl chloride is described in Chapter 3 (see p. 15).

Uses: The main use of CH_3Cl is in the production of silicones by the Rochow synthesis (see p. 179). In organic chemistry, it is used as a methylating agent giving ethers of phenols, alcohols, and cellulose (methyl cellulose). Other uses are in Friedel – Crafts – reactions to give alkylbenzenes, in the production of quarternary ammonium salts and methylmercaptan.

End products are such as tensides and pharmaceuticals.

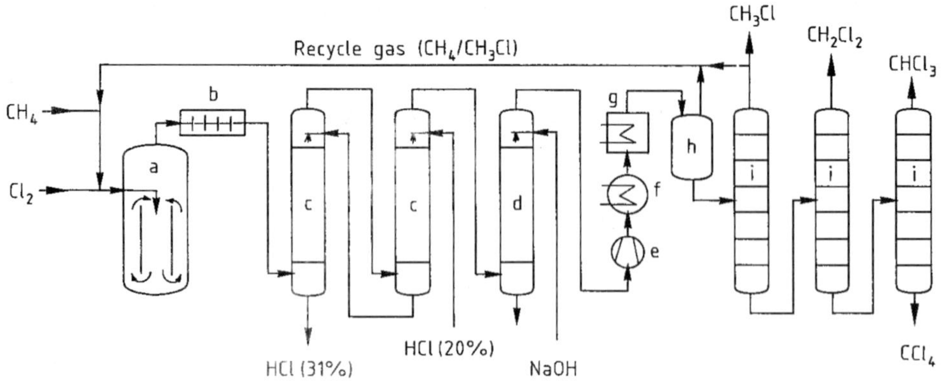

Figure 98. Methane chlorination by the Hoechst method (production of dichloromethane and trichloromethane) a) Loop reactor; b) Process gas cooler; c) HCl absorption; d) Neutralization system; e) Compressor; f) First condensation step (water); g) Gas drying system; h) Second condensation system and crude product storage vessel (brine); i) Distillation columns for CH_3Cl, CH_2Cl_2, and $CHCl_2$

In some plants, the CH_3Cl from the thermal chlorination process is recycled into the chlorination reactor to increase the yield of the higher chlorinated chloromethanes.

14.6.1.2. Dichloromethane, Methylene Chloride, CH_2Cl_2

The industrial synthesis of dichloromethane by thermal chlorination also leads to trichloromethane and tetrachloromethane. An optimal yield of this product is obtained by a large excess of methane and/or methyl chloride relative to chlorine.

The reaction (Hoechst method) is conducted adiabatically at 350 – 450 °C. The gas mixture is cooled in a heat exchanger, then the HCl is removed by washing the gas in three stages with dilute hydrochloric acid, with water and with sodium hydroxide solution or by adiabatic absorption, then dried with concentrated sulfuric acid, and finally condensed. Afterwards the products are separated by distillation. A typical output is 70 % CH_2Cl_2, 27 % $CHCl_3$ and 3 % CCl_4 (see Fig. 98).

Another process (Stauffer Chem. Co.) uses an integrated chlorination/hydrochlorination facility with methanol as raw material, in this way avoiding the complicated removal of hydrogen chloride [354].

Dichloromethane is used as a cleaning agent (e.g. in electronics) and paint remover (45 – 50 %), in aerosol formulations (20 – 25 %), as extracting agent for decaffeinating of coffee, extracting of hops, paraffin extraction, and as a solvent.

As a consequence of environmental protection measures the use of dichloromethane was declining. This had been achieved by improving production and application technologies, handling in safety vessels, and by recycling, work-up and reuse of used solvents. In Germany, the demand had been reduced by 75 % between 1986 and 1997. Since 1992 sales have stabilized at ca. 150 Kt/yr.

The effect of CH_2Cl_2 on human health and on the environment is reported in literature [355], [356].

14.6.1.3. Trichloromethane, Chloroform, $CHCl_3$

Production: see Chapter 14.6.1.2.

The principal application of chloroform is the the production of the refrigerant monochlordifluormethane, HCFC 22, $CHClF_2$, and other chlorofluoroalkanes. Because of the ozone depletion potential of chlorofluorocarbons (CFC), their open use had been restricted for some years and afterwards prohibited by the Montreal Protocol in 1987 and subsequent international meetings. The role of CFC in stratospheric chemistry is discussed in detail in [357]. HCFCs are less active, they will be phased out about 2005.

The voluntary renunciation of CFCs by many producers led to a decline also in the demand for $CHCl_3$ (see Chapter 14.9.1).

Chloroform is a precursor of tetrafluoroethene (TFE), and the demand is rising in this sector. The fluorination takes place by the exchange of the chlorine with hydrogen fluoride

$$R\text{-}Cl + HF \longrightarrow R\text{-}F + HCl$$

mostly in liquid-phase reactions with antimony(III) and antimony(V) halides as catalysts. The TFE is polymerized to give materials with exceptional thermal and chemical properties, such as polytetrafluoroethene (PTFE) and the families of fluoroplastics and fluoroelastomers [358].

To a smaller extent, chloroform is used as an intermediate for orthoformic esters and as an extractant for pharmaceutical products.

14.6.1.4. Tetrachloromethane, CCl_4

Thermal chlorination of methane is the most common process for producing CCl_4.

The perchlorination process (chlorinolysis = Chlorination and pyrolysis) uses the high-temperature chlorination of methane and the chlorinating cleavage of hydrocarbons and their chlorinated derivatives to produce CCl_4. In this way, chlorine containing by-products and the residues from other chlorination processes, such as derived from the production of vinyl chloride or of allyl chloride can be converted to a useful product.

Other (older) processes start from carbon disulfide

$$CS_2 + 2\,Cl_2 \longrightarrow CCl_4 + 2\,S,$$

from elemental carbon, or from mixtures of propane and propene.

CCl_4 had been used for the production of the chlorofluorocarbons trichlorofluoromethane, CFC11, dichlorodifluoromethane, CFC12, chlorotrifluoromethane, CFC13, as a degreasing and cleaning agent, as an absorbent in chlorine manufacture and as a solvent.

Because of its ozone depletion potential, these emissive uses as solvents have been banned [359].

The importance of CCl_4 as a chlorine consumer has decreased, also the chlorinolysis processes have been discontinued in many countries.

CCl_4 ist still used as a process agent and as chemical intermediate.

14.6.2. Chloroethanes [360]

The chloroethanes comprises the following compounds:

Monochloroethane, C_2H_5Cl
1,1-Dichloroethane, CH_3-$CHCl_2$
1,2-Dichloroethane, EDC, CH_2Cl-CH_2Cl
1,1,1-Trichloroethane, CH_3-CCl_3
1,1,2-Trichloroethane, CH_2Cl-$CHCl_2$
1,1,1,2-Tetrachloroethane, CH_2Cl-CCl_3
1,1,2,2-Tetrachloroethane, $CHCl_2$-$CHCl_2$
Pentachloroethane, $CHCl_2$-CCl_3
Hexachloroethane, CCl_3-CCl_3

EDC is by far the commercially most important of these compounds. About 30 % of the total chlorine production is used in the manufacture of EDC, mostly to obtain the vinylchloride and from this the polyvinylchloride, PVC. Ethylchloride, 1,1 dichloroethane, and 1,1,1-trichlorethane are produced industrially on an much smaller scale.

Basic feedstocks of all chloroethanes are ethane or ethene and chlorine.

14.6.2.1. Monochloroethane, Ethylchloride, C_2H_3Cl

Ethylchloride had been used as an starting material for the production of tetraethyl lead. The trend toward unleaded gasoline has reduced its commercial significance.

The technical production is carried out by the hydrochlorination of ethene

$$C_2H_2 + HCl \longrightarrow C_2H_3Cl$$

or by thermal chlorination of ethane

$$C_2H_6 + Cl_2 \longrightarrow C_2H_3Cl + HCl$$

In the production of EDC by the oxychlorination process, ethylchloride is a major byproduct. It can be recovered from the light vent gases of this process by condensation or scrubbing.

Ethylchloride is used as an ethylating agent in the production of ethylcellulose and of fine chemicals and in extraction processes e.g. for the isolation of sensitive natural fragrances.

14.6.2.2. 1,1-Dichloroethane

1,1-Dichloroethane is used as a precursor material for the production of the trichloroethanes. From numerous production routes, the catalytical addition of HCl to vinylchloride is industrially used

$$CH_2{=}CHCl + HCl \longrightarrow CH_3\text{-}CHCl_2$$

14.6.2.3. 1,2-Dichloroethane, EDC, $C_2H_4Cl_2$

The EDC belongs to the chemicals with the highest production rates, the annual production was about 14 million tonnes in 1998, the demand is still rising. It is the biggest single consumer for chlorine.

Industrially it is produced by the chlorination of ethene, either by the direct chlorination process using chlorine or by the oxychlorination with hydrogen chloride as chlorinating agent.

In practice, both processes are carried out together and in parallel because most EDC plants are connected to vinyl chloride monomer (VCM) units, and the oxychlorination process is used to balance the hydrogen chloride from the VCM production.

In the direct chlorination process ethene and chlorine are reacted most commonly in the liquid phase of 1,2-dichloroethane in the presence of a catalyst, primarily iron(III) chloride.

$$CH_2{=}CH_2 + Cl_2 \longrightarrow CH_2Cl\text{-}CH_2Cl$$

In the low-temperature chlorination the heat of the exothermic reaction is removed by indirect cooling of the reactor to temperatures below the b.p. of EDC, e.g.20 – 70 °C. The advantages are low formation of byproducts and less problems with construction materials, the disadvantage is the high energy input for the rectification of the EDC.

In the high-temperature process the reaction heat is used to distill the EDC. Special catalyst systems lower the formation of unwanted trichloroethanes. In addition, liquid raw EDC from the OXY-EDC-process can be added, since the excess heat allows the vaporization. A simplified flow sheet for the high-temperature process is shown in Fig. 99.

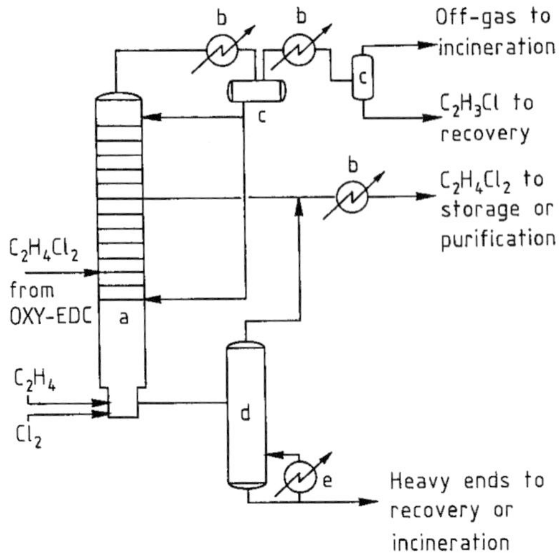

Figure 99. Simplified DC–HTC process
a) Reactor; b) Cooler; c) Knock-out drum; d) Heavy-end tower; e) Reboiler

In the oxychlorination process ethene and hydrogen chloride are reacted with oxygen in the presence of a catalyst, which consists in most cases of copper salts, the temperatures usually are above 200 °C.

The overall reaction is

$$C_2H_4 + 2\ HCl + 1/2\ O_2 \longrightarrow C_2H_4Cl_2 + H_2O$$

The source of hydrogen chloride is the cracking of the EDC in the VCM production, the source of oxygen may be air or pure oxygen. In the latter case, the volume of the waste gas stream to be treated is minimized.

The reaction takes place in fluidized bed reactors or fixed bed reactors. In a typical process, ethene and HCl are preheated and fed with air or oxygen to the reactor. The hot reaction gases are quenched and the resulting hydrochloric acid is treated as a waste water stream or cleaned by stripping for use, e.g. in a chlor-alkali electrolysis plant. After the quenching the gases are cooled indirectly, and the organic phase is washed with dilute NaOH in order to remove chloral. The off-gas is either vented after condensation and/or scrubbing or adsorption steps for air-based systems, or recycled, if pure oxygen is used. The wet EDC is dried by azeotropic distillation.

The numerous problems connected to the oxychlorination process, e.g. choosing the optimal catalyst system and grain size distribution, the selection of materials in contact with the hot, wet reaction gases containing chlorine, hydrochloric acid and solvents, removing the catalyst particles from the reaction gas and returning them to the reaction zone, removal of the produced water, recovery of unreacted ethene, have created a great variety of different technical solutions [360].

Uses: Between 85 and 95 % of all produced EDC is used for the production of vinyl chloride. Other uses are the production of 1,1,1-trichloroethane, trichloroethene, tetrachloroethene, of ethylenediamine, the synthesis of glycols, as an extracting agent for fats and oils and as a solvent.

14.6.2.4. 1,1,1-Trichloroethane

1,1,1-Trichloroethane is produced by photochemical chlorination of 1,1-dichloroethane, which is obtaind by the addition of HCl to vinylchloride in presence of a catalyst

$CH_2{=}CHCl + HCl \longrightarrow CH_3\text{-}CHCl_2$

$CH_3\text{-}CHCl_2 + Cl_2 \longrightarrow CH_3\text{-}CCl_3 + HCl$

The direct synthesis by chlorination of ethane is also employed

$CH_3\text{-}CH_3 + 3Cl_2 \longrightarrow CH_3\text{-}CCl_3 + 3\ HCl$

Uses: 1,1,1-trichloroethane was used as a solvent in numerous industrial applications such as cold and hot cleaning, vapor degreasing and in textile processing and dry cleaning. Fluorochlorohydrocarbons were made from it. The world capacity in 1984 amounted to 600 000 tonnes per year.

But 1,1,1-trichloroethane and its fluorochlorohydrocarbon derivates have an ozone depletion potential, therefore the production ceased for uses which might lead to diffuse emissions by end of 1995 under regulations of the European Union [356]. Today, its role as a consumer for chlorine is low, but it is still used in production of HCFC 1415 and 1426.

14.6.3. Chloroethenes [361]

The chloroethenes comprise:

Monochloroethene, Vinylchloride, VCM
1,1-Dichloroethene, Vinylidene Chloride, VDC
1,2-Dichloroethenes (cis and trans)
Trichloroethene, TRI
Tetrachloroethene, Perchloroethylene, PER

The 1,2-dichloroethenes are commercially unimportant, because they do not polymerize. Obtained as byproducts from the manufacture of other chlorinated hydrocarbons, they are used as feed stock for the synthesis of tri- or perchloroethenes.

14.6.3.1. Vinylchloride, VCM

VCM is one of the world's most important commodity chemicals. 1984 the worldwide consumption was about 12 – 15 million tonnes per year. In 1998 the production exceeded 20 million tonnes.

The production of VCM in industry is mainly based on the thermal decomposition (cracking) of 1,2-dichloroethane in a noncatalytical gas phase reaction

$$CH_2Cl\text{-}CH_2Cl \longrightarrow CH_2{=}CHCl + HCl$$

This process is endothermic, H = 100.2 kJ/mol.

The reaction occurs via a first order free radical chain mechanism. The crack reaction temperature is 500 – 550 °C, the pressure is mostly kept at 2.0 – 3.0 MPa. The crack furnace is of plug-flow design with tubes being placed in the convection zone of the furnace, which is equipped with a burner.

The EDC feed is evaporated at ca. 200 °C in the upper, cooler part of the furnace, the cracking takes place in the lower, hotter section. Chromium-nickel alloys are the best construction materials. The reaction gases leaving the furnace contain VCM, hydrogen chloride, and EDC as main constituents and numerous byproducts in gaseous phase as well as tars and coke. After cooling and quenching the hydrogen chloride is removed from the gas mixture by distillation and sent back to the oxychlorination process. Vinyl chloride is distilled in a second tower and drawn off as the head product. From the bottoms of the VC-tower the low-boiling impurities are removed firstly by distillation, then unreacted EDC is separated from the heavy ends in a last tower and recycled to the cracker.

Modern production of VCM takes place in an integrated, balanced process, comprising the three units

Direct chlorination
Oxychlorination
EDC cracking

Figure 100 shows such a process:

The basic feedstocks ethene and chlorine are reacted in the direct chlorination unit to give 1,2-dichloroethane. Additional EDC is produced in the oxychlorination process. The combined streams are purified and fed to the cracking unit, where the vinyl chloride is obtained. The HCl formed during the thermal decomposition is recycled to the oxychlorination process.

Advantages of this combination are low total energy consumption due to the fact that energy consuming steps (cracker and distillations) are combined with exothermic steps (chlorinations), the ability to balance the chlorine distribution, if other chlorinating processes are on site, and to convert the byproduct HCl back into the desired products EDC and VCM.

Before ethene from naphtha crackers was readily available, VCM was produced by hydrochlorination of acetylene

$$C_2H_2 + HCl \longrightarrow CH_2{=}CHCl$$

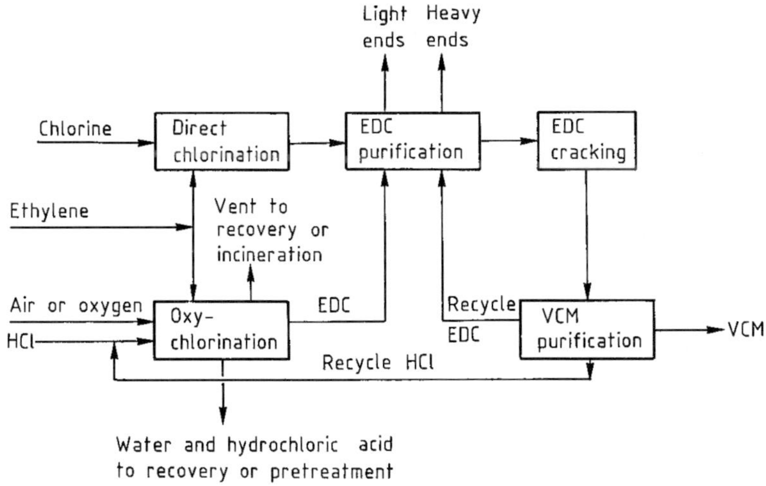

Figure 100. Ethylene-based integrated balanced process for the production of vinyl chloride

The acetylene (ethine) is obtained by hydrating calcium carbide. The production of calcium carbide consumes a considerable high amount of energy, so this process is unfavorable compared to the EDC-cracking process. Nevertheless this process may be competitive in countries with cheap coal e.g. in South Africa, and/or where ethene is not available.

Uses: Almost all of the world production of VCM is used for the production of poly(vinyl chloride), PVC [362].

PVC is a polymer prepared from the vinyl chloride monomer

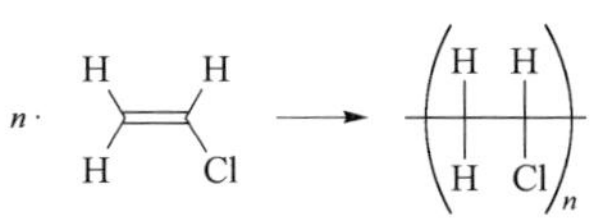

where n = 700 – 1500.

The preparation is performed by suspension, bulk, and emulsion polymerization.

PVC offers a unique combination of properties. The great variety of its properties is achieved by the different polymerization processes, copolymerization with other polymerisates, and the use of processing aids, heat stabilizers, UV-stabilizers, lubricants, plasticizers, fillers, pigments, impact modifiers, postchlorination and others.

In Table 25 a list of some typical end uses is given. The construction sector is dominating (Fig. 101) [363].

PVC is by far the biggest chlorine containing end product. In 1998 the world capacity was nearly 30 million tonnes per year. The demand for PVC is rising at a rate of 1.5 – 5 % per year, so in many parts of the world (China, India, South East Asia) new electrolysis plants are erected to cover these requirements.

Table 25. Typical end uses for PVC

Application	Rigid PVC	Flexible PVC
Construction	window frames, gutters, pipes, cars, house-siding, ports, roofing	waterproof membranes, cable insulation, roof lining, greenhouses
Domestic	curtain rails, drawer sides, laminates, audio and video tape cases, records	flooring, wallcoverings, shower curtains, leathercloth, hosepipes
Packaging	bottles, blister packs, transparent packs and punnets	cling film
Transport	car seat backs	underseal, roof linings, leathercloth upholstery, wiring insulation, window seals, decorative trim
Medical		oxygen tents, bags and tubing for blood transfusions, drips and dialysis liquids
Clothing	safety equipment	waterproofs for fishermen and emergency services, life-jackets, shoes, wellington boots, aprons and baby pants
Others	floppy-disk covers, credit cards	conveyor belts, inflatables, sports goods, toys

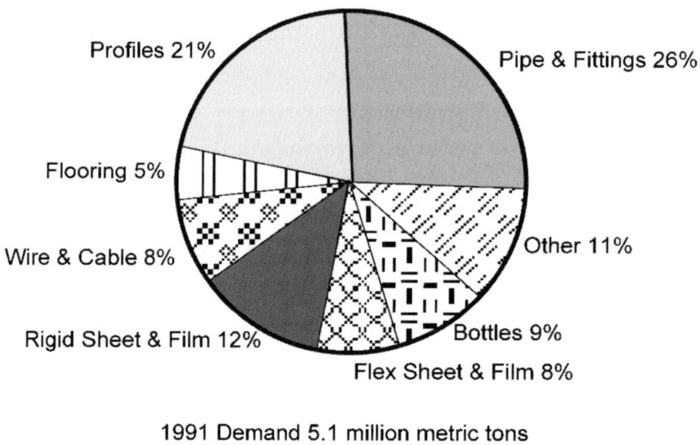

Figure 101. West European PVC demand by end use, 1991 — similar dominance by the construction sector [363]

VCM is a human carcinogen, but PVC is not. Therefore the concentration of VCM in PVC is limited by laws in many countries, e.g. in the United States (FDA 1986) to < 0.005 ppm in rigid PVC and to < 0.01 ppm in plasticized PVC, in Europe (EEC 1979) to 1 ppm in articles, and to 0.01 ppm in materials which comes in contact with food.

14.6.3.2. 1,1-Dichloroethene, Vinylidene chloride, VDC

VDC is used for the production of poly(vinyliden chloride), PVDC and its copolymers with vinylchloride, acetonitrile, methacrylonitrile, and methacrylate. These materials are used as barrier materials in the food packaging industry, for coatings on different substrates or for shrinking foils.

1,1-Dichloroethene is prepared mostly by the liquid-phase dehydrochlorination of 1,1,2-trichloroethane in the presence of alkali, e.g. NaOH

$CHCl_2\text{-}CH_2Cl + NaOH \longrightarrow CCl_2{=}CH_2 + NaCl + H_2O$

Production, properties and applications of poly(vinylidene chloride) are described in [364].

14.6.3.3. Trichloroethene, TRI

One route to trichloroethene is the dehydrochlorination of 1,1,2,2-tetrachloroethane

$CHCl_2\text{-}CHCl_2 \longrightarrow CHCl{=}CCl_2 + HCl$

in the gas-phase at temperatures of 250 – 400 °C. The reaction is strongly endothermic. In order to reduce the formation of byproducts, the reaction is carried out in the presence of a catalyst. The feedstock 1,1,2,2-tetrachloroethane is obtained by the chlorination of acetylene

$C_2H_2 + 2\ Cl_2 \longrightarrow CHCl_2\text{-}CHCl_2$

Another route to TRI starts from 1,2-dichlorethane, which is chlorinated-dehydrochlorinated in a fluidized bed reactor at temperatures between 200 – 500 °C in the presence of a catalyst.

The reaction gases are separated and purified by distillation and azeotropic distillation.

Uses: The major use for TRI is as a solvent for degreasing in the metal industry. Before use it is stabilized with acid acceptors such as amines, alcohols, epoxides and metal stabilizers. It is used also for degreasing in the textile industry, as an extractant, in solvent formulations for rubbers, paintstrippers and in paints, as intermediate for the production of dyes, colors, of fluorochloro-hydrocarbons and of fluoro-hydrocarbons.

TRI has no ozone depletion potential and no global warming potential, the toxicity is comparatively low. TRI is not suspected as a human carcinogen [365].

Before 1986 the degreasing of metals was performed in open systems without recycling of used solvent. Nowadays, closed systems for production, transport and application are applied, which prevent emissions. Spent solvents are recycled to the manufacturer, worked up and reused.

In Germany, the emissions of TRI were lowered by 95 % between 1986 and 1996. The demand for fresh TRI was 30 000 tonnes in 1986, it was reduced to 6700 tonnes in 1996 [350].

14.6.3.4. Tetrachloroethene, PER

PER is the most stable derivative of all chlorinated ethanes and ethenes. Commercially, it is the most important dry-cleaning solvent.

In industrial scale PER is produced from ethene and/or 1,2-dichloroethane through oxychlorination

$$CH_2{=}CH_2 + CH_2Cl\text{-}CH_2Cl + 2.5\ Cl_2 + 1.75\ O_2 \longrightarrow CHCl{=}CCl_2 + CCl_2{=}CCl_2 + 3.5\ H_2O$$

With this process trichloroethene and tetrachloroethene are produced simultaneously.

An interesting route to PER is the so-called "chlorinolysis" (= chlorination and pyrolysis) of $C_1 - C_3$ – hydrocarbons or chlorinated hydrocarbons through high-temperature chlorination. This process allows the preparation of PER from the residues of other processes, e.g. from vinyl chloride, allyl chloride and even oxygenated compounds.

The preferred low-pressure chlorinolysis is carried out at temperatures between 600 – 800 °C and pressures of 0.2 – 1.0 MPa. The feedstock is chlorinated and pyrolized simultaneously.

Kinetically the reaction consists of a whole series of radical crack- and substitution processes which lead to the most stable products [361]. The hot reaction gases are quenched with condensed reaction gases, then the PER is separated by distillation.

Uses: The major use of PER is as a solvent for dry cleaning. It has replaced almost all other solvents because it is non-flammable and its handling is safe [366]. The concentrations of stabilizers can be kept low. Other uses are metal degreasing (especially aluminum parts), textile finishing, dying and extracting processes. Trichloroacetic acid and some fluorocarbons are made from PER.

Since 1986, the production of fresh PER has declined rapidly for the same reasons as of TRI. Closed systems in production, transport and use, as well as recycling of used solvents, work-up and re-use) led to drastically reduced emissions.

14.6.3.5. Chlorohydrin

Chlorohydrin as a precursor of ethylene oxide has been replaced by the direct oxidation route to ethylene oxide. Its production is commercially not important.

14.6.4. Other Chlorinated C_2-Compounds

14.6.4.1. Chloracetic Acids [367]

Monochloroacetic acid, $CH_2Cl\text{-}COOH$, is produced by the hydrolysis of 1,1,2-trichloroethene, catalyzed with sulfuric acid (75 %) at 130 – 140 °C

$$CHCl{=}CCl_2 + 2H_2O \longrightarrow CH_2Cl\text{-}COOH + 2HCl$$

or by the catalyzed chlorination of acetic acid with chlorine

$$CH_3\text{-}COOH + Cl_2 \longrightarrow CH_2Cl\text{-}COOH + HCl$$

The separation of monochloroacetic acid from higher chlorinated acetic acids is achieved by crystallisation.

Monochloroacetic acid is mostly used to manufacture carboxymethyl cellulose, CMC, herbicides and thioglycolic acid. Useful derivatives are sodium chloroacetate, chloroacetyl chloride, $CH_2Cl\text{-}COCl$, to make adrenalin, chloroacetic acid esters to make chloroacetamide. The latter is a precursor for thioglycolic acid ester, vitamin A, and crop protection biocides.

Dichloroacetic acid, $CHCl_2\text{-}COOH$, is produced by hydrolyzing dichloroacetyl chloride, which is obtained by the oxydation of trichloroethane. Dichloroacetic acid and its derivatives (esters) are intermediates in organic synthesis for pharmaceuticals, antibiotics and crop protection agents.

Trichloroacetic acid, $CCl_3\text{-}COOH$, is produced by chlorination of acetic acid at 140 – 160 °C and purified by crystallization. The sodium salt is used as herbicide.

14.6.4.2. Chloroacetaldehydes [368]

Monochloroacetaldehyde, $CH_2Cl\text{-}CHO$, is very reactive. The aldehyde group takes part in reactions as well as the methylchloride group. Commercially it is produced by the reaction of chlorine with vinylacetate in water

$$CH_2{=}CHOOCCH_3 + Cl_2 \longrightarrow CH_2Cl\text{-}CHO + HCl + CH_3COOH$$

The substance and its bisulfite compounds are used as pesticides because of their bactericidal, fungicidal, algicidal, and nematicidal properties.

Dichloroacetaldehyde, $CHCl_2\text{-}CHO$, is a precursor for insecticides, diuretics and cytostatics. It is obtained by the chlorination of acetaldehyde with phosphoric acid or antimony trichloride as catalyst. Together with mono- and trichloroacetaldehyde it is formed in the preparation of acetaldehyde by the Wacker-process.

Trichloroacetaldehyde, chloral, 2,2,2-trichloroethanal, CCl_3-CHO, produced by the chlorination of acetaldehyde in hydrochloric acid, is a precursor of the famous insecticide DDT. After the ban of producing DDT in many countries, the commercial importance has declined.

14.6.4.3. Ethenechlorohydrin

Ethenechlorohydrin had been an important industrial precursor for ethylene oxide. It was produced by reacting ethene with hypochlorous acid

$$C_2H_4 + HOCl \longrightarrow CH_2Cl\text{-}CH_2OH$$

The ethenechlorohydrin was then dehydrochlorinated by means of alkali to give the ethylene oxide

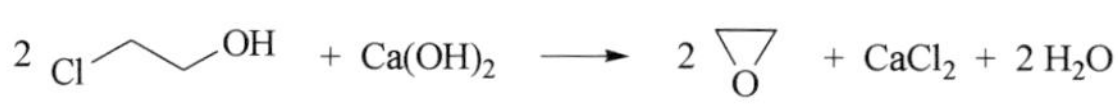

Since the introduction of the direct ethene oxidation process, the chlorohydrin route lost its commercial significance.

14.6.5. Chloropropanes [369]

From all the chloropropanes only the 1,2-dichloropropane is used on an industrial scale as an intermediate for the synthesis of perchloroethene. It is also a solvent for fats, oils, resins and lacquers and it is used for the production of roofing paper, insolation material, shoe-polish and in extraction, cleaning, degreasing and dewaxing processes.

1,2-dichloropropane is not produced directly, but is obtained as a byproduct in the synthesis of propylene oxide by the hydrin route and of allyl chloride.

14.6.6. Chloropropenes and Derivates, Propylene Oxide

14.6.6.1. Chloropropenes and Derivates

3-Chloropropene, allyl chloride, CH_2=CH-CH_2Cl, is very reactive both as an organic halide and as an olefin and is, therefore, a starting material for further synthesis. Derivates of commercial importance are allyl alcohol, epichlorohydrin, and glycerol [370].

Allyl chloride is synthesized directly by chlorinating propene under conditions favoring the substitution of hydrogen on the saturated carbon

$$CH_2{=}CH\text{-}CH_3 + Cl_2 \longrightarrow CH_2{=}CH\text{-}CH_2Cl + HCl$$

at temperatures of ca 500 °C by a radical chain mechanism. The dry feeds with propene in excess are preheated, then fed into the reactor. Rapid mixing lowers the formation of byproducts. The reactor gases are cooled and the 3-chloropropene is separated from HCl and unreacted propene and purified by distillation. HCl is removed from the propene, which is recycled into the reactor.

Uses: From 3-chloropropene mainly epichlorohydrin (90 % of the use), glycerol, and allyl alcohol are prepared. The world production in 1997 was 800 000 t allyl chloride.

Allyl alcohol, CH_2=CH-CH_2OH, is obtained from allyl chloride by alkaline hydrolysis. Oxydation of allyl alcohol gives acrolein, glycerol, formic acid, diallyl phthalate, acrylic acid. Allyl halogenids, allyl esters, allyl amines, allyl Grignard reagent can be obtained. End products are plastics, resins, plasticizers, varnish ingredients, pharmaceuticals, perfumes, flavors.

Epichlorohydrin, 1-chloro-2,3-epoxipropan [371], [372], is made from 3-chloropropen by reacting it with chlorine in aqueous phase (hypochlorination) to give a mixture of the two propylene dichlorohydrins: 1,3-dichloro-2-propanol and 2,3-dichloro-1-propanol

$$CH_2{=}CH\text{-}CH_2Cl + H_2O + Cl_2 \longrightarrow CH_2Cl\text{-}CHOH\text{-}CH_2Cl + CH_2OH\text{-}CHCl\text{-}CH_2Cl$$

The dehydrochlorination reaction with sodium or calcium hydroxide gives epichlorohydrin, salt and water

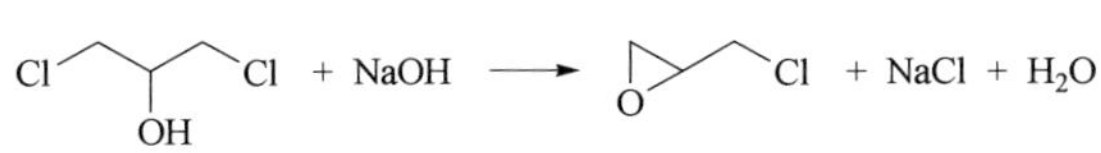

Epichlorohydrin is mainly used for the manufacture of epoxy resins (mostly by condensation reactions of epichlorohydrin and bisphenol A) [373], of glycerol ethers and glycidyl ethers, as an intermediate for organic synthesis, as disinfectant, as stabilizer of insecticides etc.

The worldwide production capacity was ca. 700 000 tonnes per year in 1993.

Glycerol, 1,2,3-propanetriol. Natural glycerol is obtained as a byproduct in the conversion of fats and oils to fatty acids or fatty acid methyl esters. Synthetic glycerol can be produced either from epichlorohydrin by hydrolysis or by hydroxylation of allyl alcohol with H_2O_2 on a WO_3-contact [374].

Glycerol is used as a raw material in the manufacure of alkyd resins, ester gums, polyurethane foams, polyols, nitroglycerine, pharmaceuticals, cosmetics, as humectant in tobacco processing, antifreeze agent, solvent, extractant, plasticizer, etc.

In 1986, the production was estimated to be ca. 550 000 tonnes/a, 75 % from the splitting of natural oils, 25 % by synthesis.

14.6.6.2. Propylene Oxide [375]

Propylene oxide, 1,2-epoxy-propane, PO, is a very reactive substance and one of the most important chemical intermediates.The worldwide propylene oxide capacity was estimated to 4.9 million t/y in 1996, the increase is ca. 4 %/y. After the PVC production, the chlorination of propylene is the second largest single chlorine consumer.

Industrially, PO is manufactured by three routes:

Chlorohydrin process
Indirect oxidation processes (Oxirane processes)
Direct oxidation process

The chlorohydrin process is carried out in two steps: the synthesis of propylene chlorohydrin, PCH, and subsequent dehydrochlorination of PCH to PO. In the first step, propene and chlorine are reacted in aqueous solution to give a mixture of 90 % 1-chloro-2-propanol and 10 % 2-chloro-1-propanol

$$CH_3\text{-}CH{=}CH_2 + HOCl \longrightarrow CH_3\text{-}CHOH\text{-}CH_2Cl\ (+\ CH_3\text{-}CHCl\text{-}CH_2OH)$$

In the second step, this mixture is dehydrochlorinated with a base, either lime ($Ca(OH)_2$) or caustic soda to form crude PO and a dilute salt stream of $CaCl_2$ or NaCl.

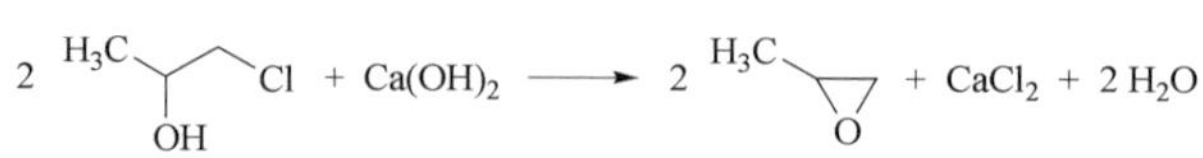

Figure 102 shows a typical PO plant using the chlorohydrin process: Gaseous propene and gaseous chlorine are mixed in roughly equimolar amounts with excess of water to form a dilute solution of the chloropropanols at 45 – 90 °C and slightly increased pressure. The reactor vent gas propane, excess propene or chlorine, oxygen, nitrogen,

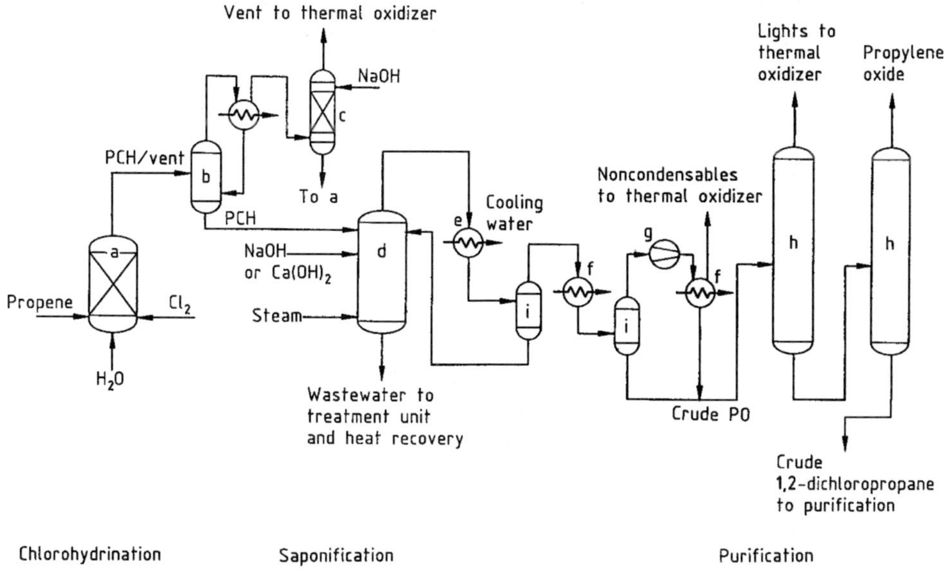

Figure 102. Typical arrangement for a propylene oxide chlorohydrin process
a) Propylene chlorohydrin reactor; b) Separator; c) Vent gas scrubber; d) Saponifier; e) Partial condenser; f) Cross exchanger; g) Compressor; h) Propylene oxide purification train; i) Drums

hydrogen, carbon dioxide, chlorinated hydrocarbons are separated from the PCH solution and fed to a scrubber. The bottoms of the separator are fed to the saponifier, where the PCH is mixed with alkali and the dehydrochlorination takes place.

The PO is removed rapidly by steam in order to prevent hydrolysis and is subsequently condensed and purified by distillation. Half of the alkali is used for the conversion of PCH to PO, the other half is required to neutralize the hydrogen chloride produced in the chlorohydrination step. Ca. 1.4 tonnes of $CaCl_2$ per tonne of PO are formed as waste water.

Byproducts are dichlorodiisopropylether, which is incinerated, and dichloropropane, which is used as a solvent.

The competitive process, the Oxirane process, starts either from isobutane or from ethylbenzene. This starting materials are converted to hydroperoxides by catalytic oxidation with air or oxygen to give tert-butyl hydroperoxide or ethylbenzene hydroperoxide. The hydroperoxides oxidize the propene in the presence of catalysts to give propylene oxide, and as byproducts either tert-butyl alcohol (2.8 t/t PO), which is converted to methyl-tert-butyl ether, or 1-phenyl-ethanol (2.5 t/t PO), which is converted to vinylbenzene (styrene).

A comparison of the chlorohydrin process with the PO/tert-butyl alcohol or PO/styrene process is difficult and depends on the integration of the raw products and the byproducts within the production site and on the product portfolio of the producing companies. Thus, a chlorohydrin PO process can be operated most economically, if it is integrated within a chlor-alkali plant. The MTBE process requires integration into a refinery complex.

Currently (1996) ca. 45 – 50 % of the world PO-production capacity is based on the chlorohydrin process, with the remainder coming from the peroxidation, the PO/styrene process covering 24 %, the PO/MTBE process also 24 %.

Uses of propylene oxide: PO is used primarily to produce polyether polyols (65 – 70 %), propylene glycols, and propylene glycol ethers.

Polyols are mainly consumed in the manufacture of polyurethanes, which are used as foams in furniture and automobile seating, bedding, carpet underlay, thermal insulation, polyethers as surface-active agents in detergents, textiles, brake fluids, and lubricants. Non-polyurethane applications include lubricants for rubber, metal rolling, drawing, antifoaming agents, deicing formulations for gasoline, hydraulic fluids.

Monopropylene glycol is prepared by the reaction of PO with water, higher propylene glycols are produced by the reaction of monopropylene glycol with PO. They are consumed as raw material for unsaturated polyester resins, as a solvent in food, drugs, cosmetics, for the formulation of coatings, plasticizers, heat transfer and hydraulic fluids, antifreezes, drilling fluids, cutting oils, and lubricants.

Propylene glycol ethers are prepared by the reaction of PO with alcohols. Typical applications are in coatings, paints, inks, resins, cleaners, waxes,and electronic circuit board lamination.

Other uses of PO are in the production of speciality chemicals, such as allyl alcohol, propylene carbonate, isopropanolamines, and hydroxypropylated cellulose.

14.6.7. Chlorobutanes [376]

1-Chlorobutane, n-butylchloride, $CH_3(CH_2)_3Cl$, is reacted with magnesium to give the correspondent Grignard reagent. It is used as an alkylating agent in Friedel-Crafts-reactions and as starting material for the production of antifouling agents in marine coatings. It is produced by esterification of n-butanol with hydrogen chloride or hydrochloric acid at 100 °C.

tert-Butyl chloride, 2-chloro-2-methyl-propane, $(CH_3)_3CCl$, is prepared by the reaction of tert-butyl alcohol with hydrogen chloride. It is used in Friedel-Crafts reactions, e.g.in the preparation of tert-butylbenzene or tert-butylphenol.

1,4-Dichlorobutane is obtained from butane-1,4-diol and hydrogen chloride in aqueous sulfuric acid at 165 °C. It serves as a synthetic intermediate, e.g. in the production of nylon.

14.6.8. Chlorobutenes [377]

Industrial and commercial important chlorobutenes are

2-Chloro-1,3-butadiene, chloroprene
1,4-Dichloro-2-butene
3,4-Dichloro-1-butene
3-Chloro-2-methyl-1-propene
2,3-Dichloro-1,3-butadiene
Hexachlorobutadiene

2-chloro-1,3-butadiene, chloroprene, CH_2=CCl-CH=CH_2, has four isomeric forms, but only the 2-chloroprene is commercially important as a monomer used to produce poly(chloroprene), known as neoprene.

There are two routes to chloroprene, a) starting from butadiene, b) starting from acetylene.

a) Butadiene is chlorinated in the vapor phase to give a mixture of 3,4-dichloro-1-butene and 1,4-dichloro-2-butene. The 1,4-dichloro-2-butene is catalytically isomerized also to 3,4dichloro-1-butene, which is then dehydrochlorinated with alkali (NaOH) to give chloroprene.
b) Acetylene dimerizes to monovinylacetylene in the presence of anhydrous copper(I)chloride and a catalyst containing alkali or ammonium salt

$$2C_2H_2 \longrightarrow CH_2{=}CH{-}C{\equiv}CH$$

The latter is reacted with hydrogen chloride to give 4-chloro-1,2-butadiene which rearranges in the presence of Cu_2Cl_2-containing catalyst to give the 2-chloro-1,3-butadiene.

Uses: The vulcanizates of poly(chloropren) have excellent resistance to weathering and ozone. Neoprene is used for electrical insulating materials, hoses, conveyor belts, flexible bellows, transmission belts, sealing materials, diving suits. Adhesive grades are used in the footwear industry. Latexes are used for dipped goods (e.g. gloves, balloons), latex foam, adhesives, fiber binders.

The production capacities in 1983 were > 650 000 tonnes per year.

1,4-Dichloro-2-butene, CH_2Cl-CH=CH-CH_2Cl , is a starting material in the production of adiponitril, butane-1,4-diol, and tetrahydrofuran. It occurs as an intermediate in the production of chloroprene. It is obtained by the chlorination of butadiene in the vapor phase at 300 – 350 °C or from 3,4-dichloro-1-butene by isomerization in the presence of Cu_2Cl_2.

2,3-Dichloro-1,3-butadiene, CH_2=CCl-CCl=CH_2, serves as a monomer for special types of poly(chloroprene). It is produced either by reacting butyne-1,4-diol with phosgene, followed by isomerization of the intermediate 1,4-dichlorobutene, or by dehydrochlorination of trichlorobutene or tetrachlorobutene with sodium hydroxide solution.

3-Chloro-2-methyl-1-propene, methallyl chloride, $CH_2{=}C(CH_3){-}CH_2Cl$, is produced by the chlorination of isobutylene in the gas phase at temperatures below 100 °C. Of greatest commercial interest are its reactions with sodium sulfite to give sodium methallyl sulfonate , and the production of 2-methyl epichlorohydrin.

Hexachlorobutadiene, $CCl_2{=}CCl{-}CCl{=}CCl_2$, is recovered as a unwanted byproduct in all chlorinolysis processes for the production of PER. A direct synthesis is not foreseen. It had been used as a pesticide, but because it is highly toxic and tends to bioaccumulate due to its low volatility, is not used today. The recovered byproduct is recycled back to the chlorinolysis or is incinerated safely.

14.6.9. Chlorinated Paraffins [378]

Chlorinated paraffins is the collective name given to industrial products prepared by chlorination of C_{10}-C_{30} paraffins and containing 20 – 70 % chlorine. Only straight-chain paraffins are used for the production of chlorinated paraffins, frequently three different mixtures: C_{10}-C_{13}, C_{14}-C_{17}, and C_{20}-C_{28}. They are prepared by introducing gaseous chlorine into the starting paraffins at 80 – 100 °C. The reaction can be initiated by visible light. Later on, the reaction heat of 150 kJ/mol is removed by cooling. The byproduct hydrogen chloride is washed out of the waste gas stream with water, the hydrochloric acid is carefully freed from organics. Batch processes are preferred, because of the large variety of special products synthesized.

The products are always mixtures of different paraffines chlorinated to varying degrees. The requirements of the customers are met by offering a range of different materials with different compositions, chlorine contents, viscosities, and stabilizers.

Uses: Chlorinated paraffins have a wide variety of applications, e.g. as plasticizers in PVC, as flameproofing agents in rubber, textiles, and plastics, as water-repellent and rot-preventing impregnants in textiles such as tents and sailcloths, as elastic sealing compounds, as components of paints and varnishes, as additive to cutting oils in metal-working,and as leather finishing to improve the resistivity against humidity.

The total world production of chlorinated paraffins is believed to be approximately 300 000 t/year, the capacity about 380 000 t.

Toxicological and environmental questions are discussed in [379]. The short-chain grades have been classified by the European producers as "dangerous for the environment", and by IMO as "Severe Marine Pollutant" and are therefore placed in UN class 9 for road/rail transport in Europe. Short-chain chlorinated paraffins were found to be carcinogenic for rodents, therefore they are classified in some countries (United States,

Germany) as suspected to be carcinogen also for humans. In a risk assessment study from Euro Chlor this relevance is questioned.

14.7. Chlorinated Aromatic Hydrocarbons

14.7.1. Nucleus-Chlorinated Aromatic Hydrocarbons [380]

14.7.1.1. Chlorinated Benzenes

The chlorinated benzenes are produces by reaction of liquid benzene with gaseous chlorine in presence of a catalyst at moderate temperatures and at atmospheric pressure. The product is a mixture of isomeric compounds with varying degrees of chlorination, the byproduct of this substitution reactions is hydrogen chloride.

1,3-dichlorobenzene, 1,3,5-trichlorobenzene and 1,2,3,5-tetrachlorobenzene have little practical use and are, therefore, avoided. The reaction is controlled by the type of catalyst, mostly Lewis acids, e.g. ferric chloride, by temperature and by the benzene:chlorine ratio in the reactor.

Batch processes are preferred because of their higher selectivity and the possibility to produce different types of chlorobenzenes in the same reactor. Continuous chlorination is used, if special chlorobenzenes are produced in greater quantities.

A mixture of chlorobenzenes, unreacted benzene, hydrogen chloride and iron catalyst leaves the reactor, and is separated by distillation.

Oxychlorination in the gas phase with hydrogen chloride and air at 240°C in presence of a catalyst is also applied.

Monochlorobenzene, C_6H_5Cl, is produced continuously, the byproduct dichlorobenzene (ca. 22 %) is separated by distillation.

The worldwide production in 1993 was ca. 365 000 t.

It is used as a solvent in chemical reactions, e.g in the production of isocyanates. Large quantities are nitrated to give nitrochlorobenzene (77 %) and subsequently converted to intermediates such as nitrophenol, nitroanisol, nitrophenetole, chloroaniline, phenylendiamine. Phenol is synthesized from monochlorobenzene by dehydrochlorination with aqueous sodium hydroxide solution at 340 °C and 28 MPa. The end products are dyes, crop protection agents, pharmaceuticals, rubber chemicals etc.

Monochlorobenzene is toxic to aquatic organisms, has a low bioaccumulation potential and a low persistence due to its volatility [381].

14.7.1.2. Dichlorobenzenes

1,2-Dichlorobenzene is an unavoidable coproduct in the monochlorobenzene production. It is used to produce disinfectants and deodorants and, after nitration to 1,2-dichloro-4-nitrobenzene, to produce dyes and pesticides.

1,3-Dichlorobenzene is applied in the production of herbicides and insecticides, pharmaceuticals and dyes. It is obtained by working-up the mother liquor from the p-dichlorobenzene crystallisation or by several direct synthesis routes.

1,4-Dichlorobenzene is used to produce disinfectant blocks and room deodorants, as moth control agent, and for the production of insecticides, of polyphenylene-sulfide-based plastics and after nitration to 2,5-dichloronitrobenzene, of dyes. It is obtained from a mixture of dichlorobenzenes by crystallisation from the melt.

1,2,4-Trichlorobenzene is used as a dye carrier and in the production of dyes, textile auxiliaries, and pesticides. It is formed in the production of mono- and dichlorobenzenes, the production rate is increased by raising the ratio chlorine:benzene up to 2.8.

Higher chlorinated benzenes, the tetra-, penta-, and hexachlorobenzenes are stable compounds with low volatility. Their production is discontinued. The use of hexachlorobenzene as an active ingredient for pesticides is legally banned in many countries, also the use of pentachlorophenol, which is a derivative of it.

14.7.1.3. Chlorinated Toluenes

The chlorotoluenes occur in five chlorination stages, mono- and tetrachlorotoluenes have three, di- and trichlorotoluenes have six isomers. Monochlorotoluenes are produced on a large scale by reacting liquid toluene with gaseous chlorine at temperatures of 20 – 70 °C and normal pressure in the presence of catalysts. Mixtures of isomers having various chlorination stages are obtained, the preferred product is monochlorotoluene. The relative proportions of the various products can be influenced by altering the catalysts and the reaction conditions. The oxychlorination with hydrogen chloride and air is also possible at elevated temperatures of 150 – 500 °C.

Monochlorotoluenes. Isomeric mixtures of the monochlorotoluenes are hydrolyzed to cresol on a considerable scale.

2-Chlorotoluene. is a starting material in the production of side-chain-chlorinated derivatives, which are precursors for dyes, pharmaceuticals, optical brighteners, and fungicides. It is also used to produce dichlorotoluenes by chlorination, 3-chlorotoluene by isomerization and o-chlorobenzonitrile by ammonoxidation.

4-Chlorotoluene is mainly used to produce p-chlorobenzotrichloride, which is a precursor for herbicides. From the relevant side-chain-chlorinated products and their derivatives pharmaceuticals, insecticides, herbicides and peroxides are made.

Dichlorotoluenes. The six isomers have very similar chemical and physical properties, they are difficult to separate from a mixture.

2,4-Dichlorotoluene and its side-chain-chlorinated derivatives are used to produce fungicides, dyes, pharmaceuticals, preservatives and peroxides as curing agents for silicones and polyesters. It is produced by the chlorination of 4-chlorotoluene.

2,6-Dichlorotoluene is a precursor for dyes and herbicides. It can be produced by chlorination of p-toluenesulfonyl chloride, followed by desulfonation, or by chlorination of 4-tert-butyltoluene, followed by dealkylation.

The other dichlorotoluenes are used only in small quantities.

2,3,6-Trichlorotoluene is also a precursor for herbicides. It can be made by chlorination of 2-chlorotoluene, of 2,3-dichlorotoluene, or of 2,6-dichlorotoluene.

14.7.1.4. Chlorophenols [382]

In chlorophenols the aromatic ring of phenol is substituted with one to five chlorine atoms. They have antimicrobial properties and are therefore used as fungicides, herbicides, insecticides, and algicides. Chlorophenols are precursors of agricultural chemicals, pharmaceuticals (e.g.clofibrate), biocides and of anthraquinone dyes.

Production: Mono-, di-, and trichlorophenols with no chlorine in a meta position are obtained by direct chlorination of melted phenol with gaseous chlorine. Tetra- and pentachlorophenol are produced batchwise by the chlorination of less chlorinated phenols in the presence of a catalyst ($AlCl_3$, $FeCl_3$). Mono-, di-, and trichlorophenols with chlorine in a meta position cannot be obtained by the chlorination of phenol, but must be prepared by other types of reactions, such as hydrolysis, sulfonation, hydrodechlorination, hydroxylation, and alkylation.

Pentachlorophenol is toxic, the decomposition rate in nature is slow, it is lipophilic and therefore liable for bioaccumulation. The use of pentachlorophenol and its salts (e.g. as wood preservation agent) is drastically restricted or banned in many industrialized countries by law, so the open use since 1976 in Europe and since 1979 in the United States, and total stop of the use since 1985 in Europe.

Chlorinated biphenyls have been widely used as insulating and cooling fluids in transformers, as dielectric impregnants for capacitors, as flame retardants, plasticizers, hydraulic fluids, etc. Polychlorinated biphenyls are chemically extremely stable and have extremely low rates of biological degradation. They are accumulating in the environment, have a tendency for bio-accumulation in fats and adipose tissue and they are considered to be carcinogens. Moreover, they give rise to the formation of polychlorinated dibenzofurans and dibenzodioxins, when they are heated to temper-

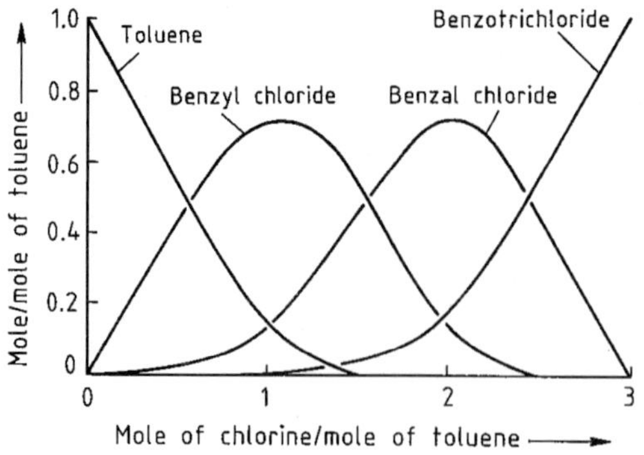

Figure 103. Progression of toluene chlorination

atures of 500 – 800 °C in presence of oxygen. Therefore, this substances are classified as POPs (Persistant Organic Pollutants). Their production, sale and use is entirely forbidden by legislation in many countries, e.g. in the United States [383], or in Western Europe [384].

Today, the polychlorinated biphenyls are recovered from the electrical devices and are safely destroyed by combustion at temperatures > 1000 °C and residence times in the oven > 2 sec.

14.7.2. Side-Chain-Chlorinated Aromatic Hydrocarbons [385]

Depending on the reaction conditions, the action of elemental chlorine can lead either to addition or substitution on the aromatic ring or to substitution in the aliphatic side-chain. The side-chain chlorinated alkyl aromatics are exceptional important intermediates for the production of numerous chemicals, including dyes, plastics, pharmaceuticals, flavors, perfumes, pesticides, catalysts, inhibitors, etc. The toluene derivatives benzyl chloride, benzal chloride and benzotrichloride have the greatest commercial significance.

The substitution of the hydrogen in the aliphatic side-chain by chlorine occurs in as a radical chain mechanism. In industrial chlorinations the formation of chlorine radicals is achieved either by irradiation (ultraviolet light, beta-radiation), or by the use of elevated temperature (100 – 200 °C). The reactants must be free of dissolved iron salts (build-up of Friedel-Crafts-catalysts), oxygen (radical scavenger), and water (build-up of hydrochloric acid).

The stage of chlorination is a function of the ratio chlorine:toluene, see Fig. 103.

Benzyl chloride is used mainly to produce plasticizers (eg. benzyl butyl phthalate), benzyl alcohol, and phenyl acetic acid via benzyl cyanide (used for the production of penicillin). Small-scale uses are for quaternary ammonium salts, benzyl esters, dyes, dibenzyl disulfide, benzyl phenols, and benzylamines.

Benzal chloride is almost exclusively used for the production of benzaldehyde. Benzal chloride is toxic and classified as a suspect carcinogen to humans (Category II-I B). These reasons require stringent measures for its handling, such as consequent sealing of the production equipment and regular medical inspection of the personnel.

Benzotrichloride is used mainly to produce benzoyl chloride by partial hydrolysis. Photochemical chlorination is widely used for the production. In order to avoid excessive chlorination and the formation of ring-chlorinated substances, cascades of six to ten reactors are applied in continuous processes. Like benzal chloride, benzotrichloride is toxic and suspected to have a carcinogenic potential.

14.8. Chlorine Balances

In the past the chemistry with chlorine was discussed mostly in an emotional manner because reliable figures were scarce or not available. In order to provide an unbiased base for fact-orientated debates with authorities and in the public, chlorine balances have been prepared since the beginning of the nineties, giving informations about the quantities of material flows as well as their structures.

The balances comprise the primary production of chlorine within an "industrial fence", follow all product chains downstream to the point of chlorinated or non-chlorinated products, take into consideration the imports and exports of chlorine, the purchases and sales of chlorinated intermediates, the recycle flows of HCl and of chlorinated hydrocarbons within the "fence", and the final disposal of chlorine containing residues. Extended balances regard also external recycle streams outside the fence, e.g. the material recycling of PVC.

The chlorine balances for Western Europe in 1992 and 1995 are described as an example [386]. The usefulness of such balances depends on the access to realistic data from competitive companies. ECOTEC in Munich, an independent consultant collected all necessary data in cooperation with the chlorine producing and using companies (21 companies with 83 production sites), checked the data and aggregated them to establish the balance.

From Figure 104 the following main findings can be derived

- The main raw material for the primary chlorine production is sodium chloride. The chlorine from electrolysis plants supplies about two third of the chlorine demand.
- One third of the total chlorine used is recycled, mainly as hydrogen chloride or hydrochloric acid.
- Only ca. 45 % of the chlorine is sold as chlorine containing products or as elemental chlorine. The leading product of this group is the PVC.
- Major non-chlorinated endproducts are propylene oxide, isocyanates and epoxy resins.

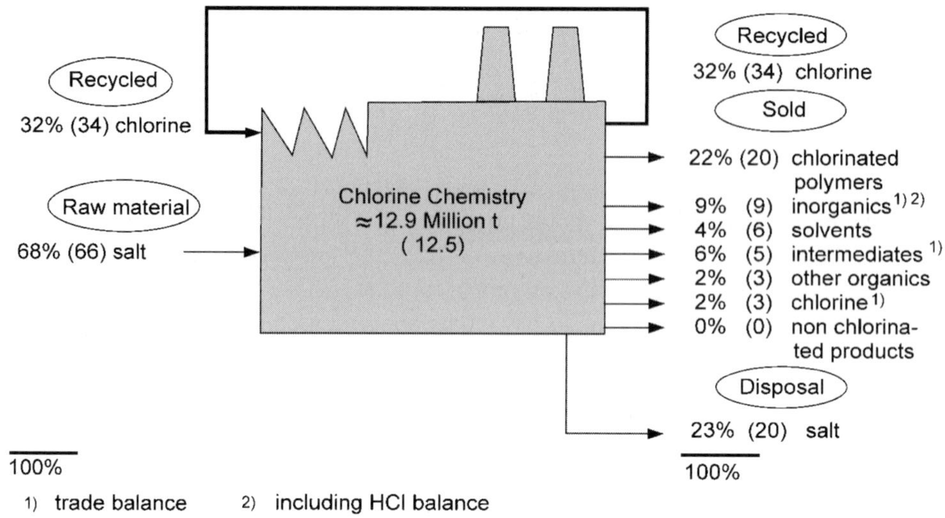

Figure 104. Chlorine flow balance 1995 (1992 flows in brackets) in Western Europe (83 locations ≙ 94 % of Western Europe)

- Ca. one fourth of the used chlorine is disposed off as residues, mainly as salts (sodium chloride and calcium chloride).

The difference between the figures of 1992 and 1995 demonstrate remarkable changes in the use pattern of chlorine:

Within the marketed chlorinated products, the importance of chlorinated polymers, particularly PVC, is increasing. This increase compensates for the decrease in the other sectors.

The demand for chlorinated solvents has dropped by 17 %, reflecting the industry's efforts to a more-efficient recovery and containment and the stricter legal regulations. Additional investigations proved, that the same volume of cleaning was achieved with less "fresh" material.

Chlorinated paraffins and derivates of chlorobenzenes are declining, as well as substances which are regulated by the Montreal Protocol. The production of 1,1,1-trichloroethane and of CFC11/12 has stopped in Europe.

The recycling of chlorinated residues by means of chlorinolysis to give PER is no longer permitted, so these residues are burnt to give hydrochloric acid.

In the production of non-chlorinated products chlorine acts as a facilitator. After separation from the product it is recycled mainly as hydrogen chloride or disposed as a salt. The production volume has increased by ca 15 %, caused mainly by the rising demand for polyurethanes, resins and particularly polycarbonates. Titanium dioxide is an example for inorganic non-chlorinated compounds.

The volume of recycled hydrogen chloride was more than 5.7 million tonnes per year. A separate balance for HCl is useful. Supplied as a by-product from chemical reactions (92 %), from the incineration of chlorinated hydrocarbons (ca. 3 %), and from direct

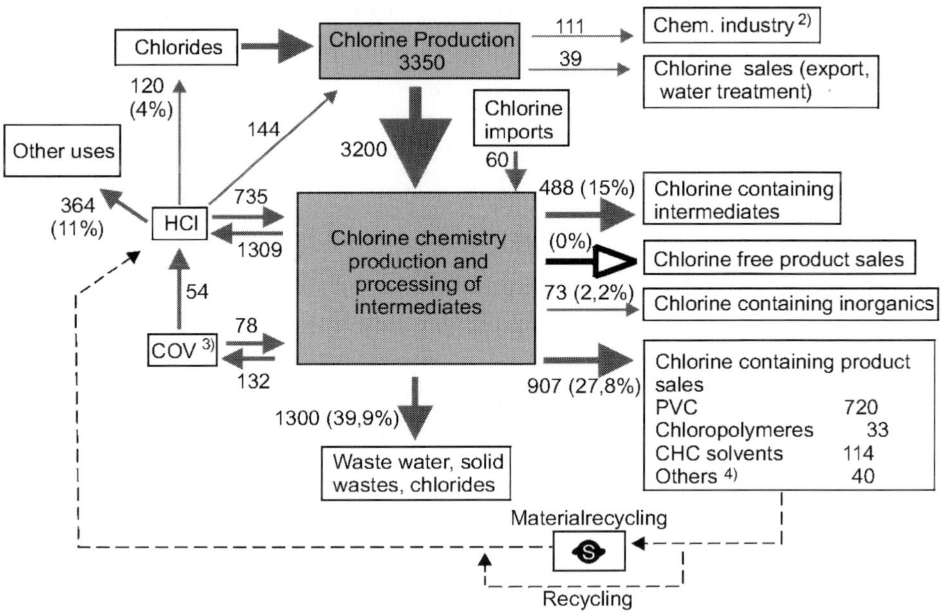

Figure 105. Chlorine flow in Germany (in kilo tonnes) in 1995
1) chlorine producers; 2) non-chlorine producers; 3) chlorinated organic compounds; 4) chlorine containing pharmaceuticals, pesticides, dyes, etc.

manufacture e.g. the production of sodium or potassium sulfate from the salts (ca. 5 %), it is consumed by recycling it into the production, e.g. into the oxychlorination process to give EDC (59 %) or other production chains (10 – 12 %), converted into chlorine by HCl-electrolysis (3 – 4 %), sold on the market (15 – 16 %), or internally used for water treatment, neutralization reactions etc (10 – 11 %).

Figure 105 shows the more detailed chlorine flow in Germany 1995 [350]. In this diagram the HCl -flows and the flows of chlorinated organic compounds are integrated. The flows of externally recycled material, such as PVC are shown. Here only 28 % of the consumed chlorine is contained in the marketed chlorinated products, 40 % of the chlorine input is disposed as wastewaters or chlorides. The decisive role of HCl in the flow management of chlorine can be seen from the fact, that the flow of chlorine as HCl is 50 % greater than the amount of chlorine in the marketed chlorinated products.

14.9. Environmental Aspects

The production and use of chlorine, as well as certain chlorinated substances bear risks for human health and the environment. International discussions are underway on legally-binding conventions and protocols thereof. Matters of concern include the use of mercury, asbestos, chemicals having an ozone-depletion potential (e.g., chlorofluorocarbons) and chlorine containing materials which are persistant, toxic, and liable to bioaccumulate (e.g., polychlorinated biphenyls, dioxins and furans, DDT, endrine, chlordane). Regarding the principles of "Responsible Care" and of "Sustainable Development" [288] the chemical industry is managing this challenge: Once a hazard is realised, the substance in question is handled individually by investigating its properties and its impact on the environment on a scientific basis. Risk assessments and risk–benefit studies are prepared, the results are published as fact sheets, position papers, scientific publications and contributions to the discussion with governments, authorities, and the public. Practical measures for reducing the risks are continuous efforts to reduce or eliminate emissions (mercury, asbestos), offer recycling concepts for products (PVC, chlorinated solvents), improve production processes and transportation safety, cease the production (CFCs, pentachlorophenol in most countries) or marketing of problematic compoundes, and to develop alternative processes (bleaching in the pulp and paper industry). All these measures are influencing the use pattern of chlorine: the use for bleaching, for C_1 compounds, for solvents, and for tetraethyllead will further decrease; the importance of organic chemicals will rise and PVC will expand especially in the construction sector. A study conducted by ECOTEC [289] illustrates the chlorine flow in Europe within the industrial production system and to external consumers.

14.9.1. Ozone Depletion and Global Warming

14.9.1.1. Ozone Depletion

In preceding chapters ("Uses") on chlorinated hydrocarbons and their derivatives it was pointed out, that the production and use of certain chlorofluorocarbons (CFC) and chlorofluoro-hydrocarbons (HCFC) was reduced or discontinued because of their influence on the ozone layer in the stratosphere. In this chapter these facts are discussed in more detail.

The ozone layer in the stratosphere filters the high-energy ultraviolet radiation out of the sunlight and in this way enables life on earth. Since midst of the seventies a drastic decrease in the ozone concentration above the antarctic region has been observed in September and October every year, and since 1990 also above the arctic region. Investigating the role of chlorine in the atmospheric chemistry it was found, that chlorofluorocarbons can reach the stratosphere, where they decompose and release halogens that affect the ozone layer [357].

Ozone is generated predominantly in the upper stratosphere by absorption of solar radiation by molecular oxygen to give oxygen atoms which combine with O_2-molecules to give ozone

$$O_2 + h\nu \longrightarrow O + O$$

$$O + O_2 \longrightarrow O_3$$

At the same time, radiation destroys ozone

$$O_3 + h\nu \longrightarrow O + O_2$$

$$O + O_3 \longrightarrow O_2 + O_2$$

thus the ozone concentration is in an equilibrium, the energy of the radiation is converted into heat.

Several reaction mechanisms are suggested on the role of halogens for the ozone destruction, e.g. chlorine acting as a catalyst

$$Cl' + O_3 \longrightarrow ClO' + O_2$$

$$ClO' + O \longrightarrow Cl' + O_2$$

$$O_3 + h\nu \longrightarrow O + O_2$$

net reaction: $2\ O_3 \longrightarrow 3\ O_2$

The destruction mechanism is a very complex cluster of reactions, influenced by the presence of stratospheric clouds, nitric and sulfuric acids, different chlorine compounds such as HOCl, HCl, Cl_2, $ClNO_3$, ClOOCl (chlorine peroxide), bromine compounds, by the sun radiation and the different temperatures in different latitudes and heights.

The role of the CFCs and HCFCs is that they could provide an important source of chlorine containing species and of chlorine free radicals to the stratosphere. They act as carriers for the transportation of chlorine atoms into the stratosphere.

Fully halogenated CFCs have been and still are to a lesser degree used as propellants for sprays, as solvents, as blowing agents for plastic foams, as refrigerants, etc. Commercially important CFCs are R11, R12, R113, R114, and R115.

For these uses they have many excellent properties: they are not toxic, not flammable, have a low thermal conductivity, they are practically insoluble in water and they are chemically very stable.

The latter two properties create the problem: Because of the insolubility in water, they are not removed by rainfall, and they are inert towards the hydroxyl radical. The reaction with this radical to form water is the process, that initiates the oxidation of hydrocarbons. Thus the CFCs are not removed by the common cleansing mechanisms that operate in the lower atmosphere, instead they rise into the stratosphere, where they are destroyed by solar short-wave UV-radiation releasing the ozone-depleting chlorine atoms. Because transport into the stratosphere is very slow, the residence time for CFC's in the environment is extremely long, up to the order of one century, so they accumulate in the atmosphere.

In order to make the efficiency of the various CFCs comparable, they are characterized by their Ozone Depletion Potential (ODP). This is an equivalence number, which is defined as the calculated ozone depletion due to the release of 1 kg of the compound, divided by the calculated ozone depletion due to the release of 1 kg of CFC11 (CCl_3F). The ODPs of the above mentioned CFC's are between 0.4 for R115 and 1.1 for tetrachloromethane.

In a similar way, the Global Warming Potential (GWP) is defined, also related to the effect of 1 kg of CFC11 [387].

Trouble-shooting: The Montreal Protocol. Under the auspices of the United Nations Environmental Programme (UNEP) the first international agreement limiting the production of CFC's was approved by 43 countries in September 1987. Initially a reduction of 50 % by the end of the century was foreseen. In view of the strength of scientific evidence in the following years, additional amendments in London (1990), Kopenhagen (1992) and Vienna (1995) strenghened these objectives: In industrialized countries, the production of CFCs was phased out at the end of 1995, and other compounds such as halons, methyl bromide, tetrachlorocarbon, and 1,1,1-trichloromethane were also regulated. The provisions of the Montreal Protocol are enforced in many countries, e.g. in Europe by the EU Council Regulation No. 3093/94 on substances that deplete the ozone layer from December 1994. The regulatory processes are still going on.

In order to enable the fast discontinuation of the production and use of fully halogenated CFCs, hydro-chlorofluorocarbons were introduced in industry as a transition stage. The atmospheric fate and impact of these hydrochlorofluorocarbons and chlorinated solvents are described in [387]. The authors come to the conclusion, that these compounds, with the exception of 1,1,1-trichloroethane, make a small or insignificant contribution to the stratospheric ozone depletion, global warming, "photochemical smog", "acid rain" or chloride and fluoride levels in precipitations. The ozone depletion potentials are 10 to 50 times lower than that of CFC11 or CFC12, mainly as a consequence of their shorter atmospheric lifetime — some months to 10 years — due to destruction in the atmosphere.

In spite of these facts, the Montreal Protocol limits the use of HCFC to the substitution of CFCs, with a phase-out before 2030. In some countries (EU) the legislation is more stringent: in order to get a stepwise reduction of the emissions of HCFCs, the use in aerosols, as solvents or as refrigerants is prohibited, but not when they are used as feedstock or as processing agent and for special defined uses.

As substitutes for the chlorine containing CFCs and HCFCs, partially fluorinated hydrocarbons such as R23, R125, R134a, R143a, and R152a are investigated. They do not contain chlorine, therefore their impact on the ozone layer is negligible.

For the chlorinated solvents methylenechloride, TRI and PER the lifetimes are much shorter than that of HCFCs or 1,1,1-trichloroethane.

As a result of this efforts, the stratospheric reactive chlorine loading had its maximum between 1996 and 1999, and it is predicted to decline over the next decades.

"The CFC-ozone depletion problem has demonstrated that humankind is capable of seriously modifying the atmosphere on a global scale, it has also shown us that in

Table 26. Contributions to radiative forcing

Compounds	% of total
CO_2	54.9
CH_4	16.5
N_2O	4.9
O_3 (tropospheric)	14.1
CFC11 and CFC12	7.0
Other CFCs	1.7
HCFCs	0.8

principle society can solve these global problems by means of international agreements." M. J. Molina in [357].

14.9.1.2. Global Warming

The visible and ultraviolet radiation from the sun is partly absorbed by atmospheric gases and partly reflected by clouds and the earth's surface, but most is absorbed by the earth. The latter re-emits energy at a much longer wave-length. This infrared radiation is absorbed to some extent by gases in the atmosphere, both natural compounds and pollutants, causing the lower atmosphere and the earth's surface to warm up slightly.

The so-called "greenhouse effect" of the contributing compounds depends to a large extent on their concentration and lifetime in the atmosphere and on their infrared absoption spectra. Fluorocarbons are efficient greenhouse gases because they absorb specifically in the infrared spectrum. The relative contributions of chlorine compounds are expressed by their Halocarbon Global Warming Potentials (HGWP), which is defined like the ODP (see above). The absolute contributions are expressed as "radiative forcing", which is the increase in infrared radiation flux towards the earth due to the presence of the gases. The main contributors to radiative forcing are given (ca. 1995) in Table 26.

Carbon dioxide, methane, tropospheric ozone, the CFCs, and nitrous oxide contribute most to the greenhouse effect. HCFCs represent less than 1%, the chlorinated solvents part is roughly estimated at 0.04% of the total.

The relative global warming potential (GWP) of some CFCs, HCFCs and of 1,1,1-trichloroethane is given in Table 27.

The phasing out of CFC11, CFC12 and CCl_4 according to the Montreal Protocol eliminates at the same time significant contributors to the global warming.

The direct effect of the emission of a greenhouse gas to the atmosphere is not sufficient to judge the impact on global warming, however the whole system should be considered. Each application consumes energy and in this way leads to a corresponding release of the greenhouse gas CO_2, resulting from the production of the energy, the so-called indirect effect.

Table 27. Relative Global Warming Potentials of some CFCs, HCFCs, and Trichloroethane

Substance	CCl_3F	CCl_2F_2	$C_2Cl_3F_3$	$C_2Cl_2F_4$	C_2ClF_5	$CHClF_2$	CCl_4	1,1,1-Tri
=	CFC11	CFC12	CFC113	CFC114	CFC115	HCFC22		
rel. GWP	1.0	2.8 – 3.4	1.3 – 1.4	3.7 – 4.1	7.4 – 7.6	0.32 – 0.37	0.34 – 0.35	0.024

The Alternative Fluorocarbons Environmental Acceptability Study (AFEAS) in collaboration with the United States Department of Energy has developed the TEWI concept (Total Eqivalent Warming Impact), which takes into consideration both of these effects. An analysis of TEWI values for systems using HCFCs or HFCs as well as for non-fluorocarbon technologies, shows that the relative importance of the direct and indirect effects varies widely depending on the type of application. For energy-intensive applications (refrigeration, air conditioning, foam blowing), the fluorocarbon systems may contribute less overall to global warming than the competitive alternative technologies, on account of the superior energy efficiency of the former [387].

14.9.2. Dioxins

On July 10th 1976 an explosion occurred in a chemical plant in Seveso, Italy, releasing some kilogramms of "dioxins" into the environment. This event and its consequences led to an emotive controversal discussion in the scientific community and in the public, which made "dioxin" a synonym for almost all problems of the chemistry with chlorine.

A description of all aspects of this discussion is beyond the scope of this book. Details can be seen in the literature [388]–[390] and the references quoted therein.

Dioxin is most common a name given to a group of molecules made of two chlorinated benzene rings connected by one or two atoms of oxygen.

The number of chlorine atoms varies from 1 to 8. The different combinations of both the number of chlorine atoms and their positions of substitution in the molecules gives a large number of isomers. The group of "dioxins" comprises 75 dioxins (polychlorinated dibenzo-p-dioxines, PCDD) and 135 related furans (polychlorinated dibenzofurans, PCDF).

The toxicity of the isomers varies considerably among the different PCDD's and PCDFs. Only about 17 out of the 210 dioxin and furan congeners are toxic. The greatest toxicity possesses the 2,3,7,8-tetra-CDD (the so-called "Seveso poison"), and the 2,3,7,8-tetra-CDF. By convention, the toxicity of a mixture of PCDD's and PCDF's is determined as "Total Toxic Equivalent = TEQ" by multiplying the concentration of the different congeners with a specific "Toxic Equivalent Factor = TEF"

$$TEQ = \sum_i (TEF_i x c_i)$$

The TEF-value for 2,3,7,8-tetra-CDD is set = 1 by definition, the same value is valid for 2,3,7,8-tetra-CDF. The factors of the other dioxins vary from 0.001 for octa-CDDs and octa-CDFs, to 0.01 for hepta-CDD's and hepta-CDFs, to 0.1 for hexa-CDDs and hexa-CDFs, and to 0.5 for penta-CDDs and penta-CDFs.

Dioxins are not manufactured as commercial products or ingredients. In chemical practice they occur as unintented byproducts of incomplete combustion and of certain chemical processes. Dioxins are formed whenever organic material and chlorine are present in an incineration process, in industry as well as in nature (e.g. forest fires, volcanoes). The formation is favored at temperatures of ca 300 °C, at temperatures above 600 °C, they begin to decompose, at temperatures above 850 °C they are destroyed. The formation seems to be independent from the source of the chlorine, e.g. whether the chlorine is contained in a salt or in PVC.

Accidental fires of chlorine containing materials are point sources for the emission of dioxins.

Several countries have performed national source inventories for dioxins to eliminate the major sources and minimize the potential risk for the environment.

The contribution of the chemical industry to the total emission of dioxin is small [391]. According to a study in The Netherlands in 1994 the share was ca. 0.1 %.

In 1995, in the United Kingdom the dominant source was the incineration of solid municipal waste, contributing an average of 70 % to the atmospheric emissions. Other major emissions are from sinter plants (steel mills), combustion of coal, emissions from iron and steel plants, from non-ferous metal operations, and combustion of clinical waste, summing up to 23 % of total industrial emissions [389].

Wood burning, the use of leaded gasoline with chlorine containing scavengers, and smoking are further sources.

Dioxins have a low vapor pressure, do not readily dissolve in water, but are lipophilic. They are chemically stable and do not react easily with other chemicals. They are adsorbed at suspended solid material in water and in sediment and at soot in case of fires. Because of these properties, they are persistent, and liable to bioaccumulate.

Human exposure to dioxins stems from the food chain (> 90 %), from the ingestion of milk, dairy products, meat and fish, all containing dioxins. The ADI-value (Acceptable Daily Intake) is in the order of 1 picogram per kilogram of body-weight.

TCDD has been shown to be carcinogenic for certain animals, therefore it is classified as suspected to be carcinogenic to humans in some countries. Although its decision continues to be controversal, the International Agency for Research on Cancer decided in 1997 to classify TCDD as carcinogenic to humans. On the other hand, epidemiological studies have not revealed that dioxins present a risk of cancer to humans. A clear human reaction to high levels of exposure is chloracne — a serious but reversible skin condition.

Trouble Shooting. Two sources of dioxins have to be distinguished: primary sources, which are responsible for the formation of dioxin, and secondary sources, from which existing dioxin is emitted.

1990, in Germany the formation of dioxin by incineration was about 1 kg TEQ per year. By optimizing the burning processes and by waste gas cleaning in industrial incineration plants, the emissions from these sources were lowered to 10 g TEQ/year.

The chemical industry has lowered its emissions by discontinuing processes, which led to the formation of dioxin, such as the production of trichloro- and pentachlorophenols. The use of these substances is legally banned. Other processes, where dioxin can be formed as a byproduct, have been altered in order to protect the personnel and to prevent emissions, for example the production of chlorophenoxy acetic acid.

In Europe, the production, storage and transportation of dangerous goods is regulated by the Council Directive 82/501/EEC of June 24, 1982 (the so-called "Seveso Directive") on the major accident hazards of certain industrial activities, and by several amendments [392]. In the same reference, the legal activities in the United States, Europe and Japan are described and compared. The corresponding laws in the United States are the Toxic Substances Control Act 1976 (15 U.S.C §§ 2601–29) and the Emergency Planning and Community Right-to-Know Act, in Japan the Chemical Substances Law.

Certain industrial processes, e.g. the recovery of copper from cables by burning the insulation or the recovery of scrap led to a considerable contamination of the soil in the neighborhood of these plants. The entry of these dioxins into the food chain must be prevented primarily by restricting the use of these grounds for the production of food, further technologies are developed to identify and reduce the amount of dioxins released into the environment as well as to treat dioxin containing materials and wastes most effectively.

The widespread use of unleaded gasoline has brought a further reduction of the emissions of dioxin.

The primary sources of dioxin have been shut down or are reduced to a very low level.

A more difficult task for the future is the lowering of emissions from diffuse sources such as private wood burning or the treatment of wood which was protected by pentachlorophenol in the past.

There is some evidence that dioxins have been present in the environment for thousands of years and have been identified as contaminants worldwide. The amount of dioxins observed rose between 1930 and 1960, peaked in the 1960s and 1970s, then started declining. The level of dioxin in the environment has continued to decrease since then.

The positive results of all the efforts can be seen from investigations of sediments, samples of which have been taken from isolated lakes around the world. Analysis data indicate that this decline could be as high as 80 % since 1970 (see Fig. 106) [388].

A French document from 1994, published under the auspices of the French Academy of Science stated that, based on the present knowledge, the population exposure to current levels of dioxins does not pose a major risk to health [393].

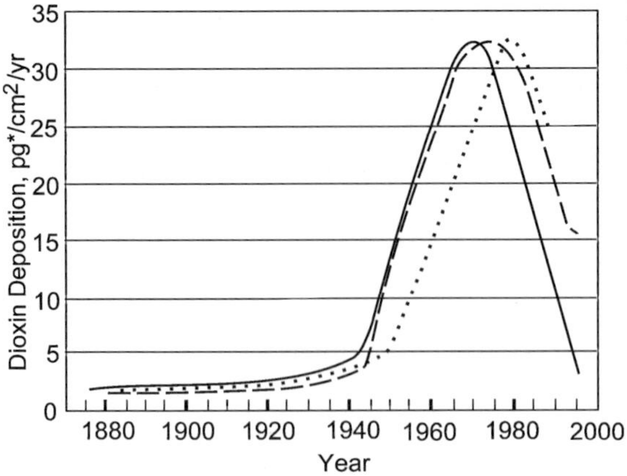

Source: Hagenmaier, 1996; Hites, 1990.
*1 Picogramme (pg) = 0.000 000 000 001 of a gramme

Figure 106. Dioxin deposition trends in the environment (from measurements in three lake sediments cores)

14.9.3. Persistent Organic Pollutants, POPs

Chemicals, which are persistent, toxic and liable to bioaccumulation, are called PTBs. They have primarily local effects. Persistence is the evidence that the substances' half-life is greater than two months in water and greater than six months in soil or sediment. Toxicity is the potential to adversely affect human health and/or the environment. Bioaccumulation is the evidence that the Bio-Accumulation Factor (BAF) is greater than 5000. Up to 1995, there was no clear definition of which products belong to this class [394]. Heavy metals, such as mercury and POPs fall into this category.

POPs are persistent organic pollutants, which persist in the environment and are toxic to humans and/or wildlife, have a strong tendency to bioaccumulation in the food chain, and are prone to long-range transport [395].

According to an UN-ECE Protocol (United Nations Economic Commission for Europe), POPs generally consist of three different groups: industrial chemical products such as PCBs, byproducts such as dioxins and furans, and pesticides such as DDT. The North American Commission on Environmental Cooperation (NACEC) has adopted a program on the management of persistent toxic substances. The United Nations Environment Program (UNEP) has begun negotiation on a legally-binding global agreement to address POPs in June, 1998.

The ca. 100 potential POPs include the following 12 chlorine-related substances: PCBs (polychlorinated biphenyls), dioxins and furans, aldrin, dieldrin, DDT, endrin, chlordane, hexachlorobenzene, Mirex, toxaphene, and heptachlor [396].

The relationship between physico-chemical properties, environmental distribution and ecotoxic effects are discussed in [389]. The environmental fate of POPs is described, including their transport and dispersion as well as their accumulation and transformation in defined environmental compartments. One of the findings is, that biotic and abiotic persistence increases with the degree of chlorination, therefore favoring long-term bioaccumulation.

The toxicity of special persistent chlorinated organic compounds, such as PCBs, dioxins, DDT is reported in [389].

Often, the same properties that make a chemical beneficial and useful can also make it hazardous if not properly managed.

Trouble Shooting. The industry actively supports and participates in international negotiations aimed at establishing global and regional agreements to reduce the risks of POPs, such as the recently completed POPs-protocol of the United Nations Economic Commission for Europe (UN-ECE). To ensure that these agreements reduce the risks of these substances without imposing unnecessary social or economic consequences, industry believes that any international action should be based on a scientific assessment of potential risks and use of a science-based precautionary approach as embodied in Principle 15 of the Rio Declaration. (World Chlorine Council (WCC), Leaflet: "Managing Persistent Organic Pollutants", Arlington, USA). There is particular need for reasonable, balanced and science-based decision-making in a product-by-product based cost-benefit analysis. In this respect, Euro Chlor is preparing risk assessment studies for all compounds enlisted in the protocols.

Most of the POPs are already banned and discontinued in Western Europe and many other countries of the world, but are still in use in developing countries.

Practical measures for the reduction of POPs in the environment are

- discontinuing the production of the substances
- discontinuing the open use (e.g. pesticides)
- alter the production process to avoid intermediates which are POPs
- improve the production process to lower the formation of by-products (e.g. dioxins) by making use of the BAT (best available techniques)
- reduce the emissions by treating the waste gas streams and the effluents.

A demonstration of the positive results of all these efforts are given in [397]. This paper shows the reduction of the emissions of 24 chlorinated substances between 1985 and 1995. In most cases the emissions into water and air have been reduced to less than 10 % of the original value.

DDT. The history of DDT shows the dilemma of managing POPs:

In 1939, P. Müller discovered the insecticidal properties of dichloro-diphenyl-trichloroethane (DDT). High insecticidal toxicity and low mammalian toxicity made it worldwide the most successful insecticide for years. P. Müller was awarded the Nobel Prize for Medicine in 1948. Within a few years the production grew to more than 100 000 tonnes per year. It was used in huge quantities against moths, lice, mosquitos, flies, etc., many of them transmitters of diseases such as malaria, typhus, cholera. These diseases could be reduced drastically.

In the 60ties, the POP-properties of DDT became known. Highly resistant to chemical and biological degradation, DDT and its metabolites (mainly DDE) were found in all parts of the world, even in human breast milk. Their biological accumulation factor in the food chain is very high (10 000 to 100 000), and they are highly lipophilic, so DDT and DDE are accumulated in fat tissues.

Because of these properties and the suspect of being carcinogenic and having estrogenic properties, the production and use of DDT is legally forbidden in most industrialized countries since the 70ties. As one consequence, the concentration of DDE in breast milk of Swedish women was lowered by 90 % by 1996 [398].

In tropical countries, the ban had dramatic effects on the health of the population. In Sri Lanka, for example, more than two million people were taken ill with malaria before 1950. After a campaign of the WHO with DDT against the anopheles mosquitos from 1961 to 1963, this disease had disappeared almost totally, so the campaign was stopped. Within the following five years the number of diseases rose to 2.5 million per year. A substitute with comparable outstanding effectivenes and low prices is not yet known. Therefore, DDT is produced and carefully used in several developing countries also today.

15. Economic Aspects

In North America (United States and Canada), the Chlorine Institute estimated the value of sales of the chlorine-dependent industries in 1990 to be ca. 80 billion \$, the total benefits at the chlorine industry to consumers ca. 120 billion \$. About 1.4 million workers were directly or indirectly employed in chlorine-dependent industries. (The Chlorine Institute, "Assessments of the Economic Benefits of Chlor-Alkali Chemicals to the United States and Canadian Economies" by Charles River Associated Inc. April 1993).

In Western Europe in 1995 [290]:

- Almost 2×10^6 jobs were related to chlorine
- 55 % of European chemical turnover depended on chlorine
- 85 % of pharmaceuticals are made using chlorine
- 98 % of Western Europe's drinking water is purified by chlorination

These figures demonstrate the economic importance of the chlorine industry.

Figure 107 shows how world chlorine production has developed from 1965 to 1995. The figures are only estimates, as many important nations do not publish this information.

Chlorine capacities 1995 (in 10^3 t/a) in the main producer countries was as follows [291]:

United States	11 860
Japan	4250
Former Soviet Union	3800
China	3750
Germany	3690
Brazil	1660
Canada	1630
France	1580
United Kingdom	1270
India	1230
Italy	1130

World chlorine capacity was about 40×10^6 t/a in 1983, it rose to 48×10^6 t/a in 1998, and is forecast to reach 55×10^6 t/a by 2005 [284].

Figure 108 demonstrates the growth of chlorine capacity by economic regions [292]. Compared to 1983, the capacities in 1997 remained nearly constant in Western Europe, Eastern Europe, and Africa. In North America, reductions in Canada were outweighed by growth in the United States. Huge capacities were erected in the Middle East and above all in Asia and Oceania (China, India), where capacities were tripled. From the growth between 1997 and 2005, North America`s share will be 35 %, Asia`s 40 %, and the Middle East's 15 % [284]. After 2000 China will become the second biggest producer after the United States, with an annual chlorine capacity of about 5.5×10^6 t.

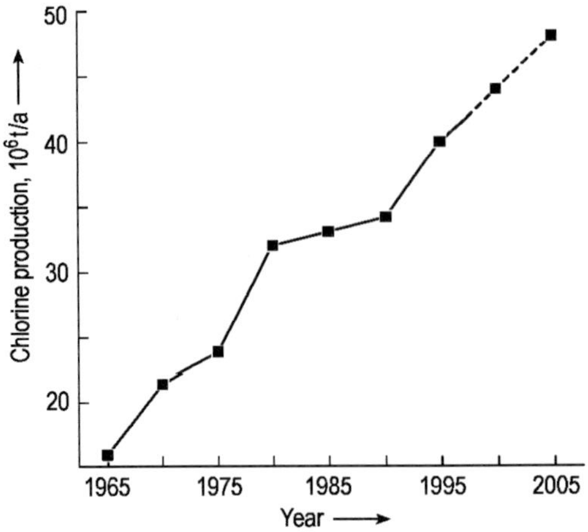

Figure 107. World chlorine production (1965 – 1995)

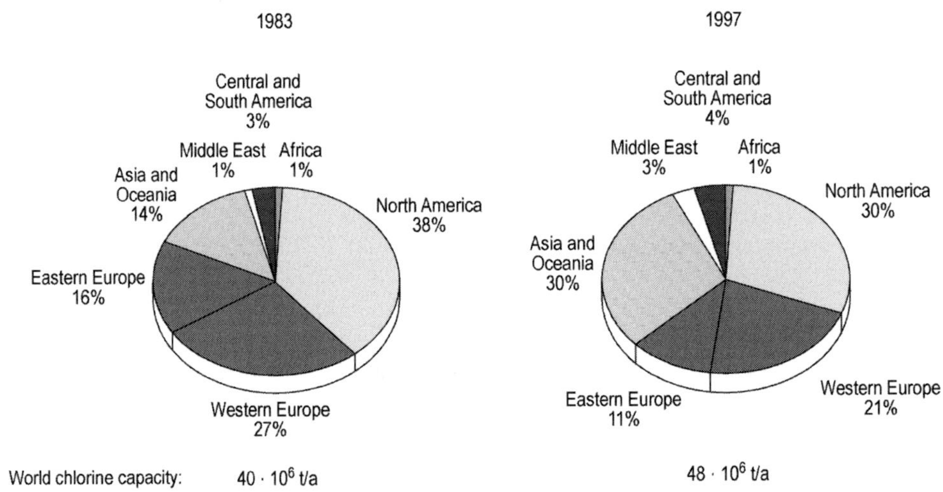

Figure 108. Chlorine production capacities by economic regions (1983 and 1997)

All chlorine producers are listed in [293]. A detailed review of the chlor-alkali market for 1994 is given in [294].

The production costs of 1 electrochemical unit (≙ 1 t chlorine + 1.13 t sodium hydroxide) depends up to 60 % on the price of electricity. At 3.5 ¢/kWh, they are about 250 \$/ecu (Fig. 109), so the total value of production is ca. 24×10^9 \$/a. The chlor-alkali industry is one of the biggest consumers of electrical energy, consuming ca. 0.15×10^{12} kWh/a.

A prediction of the market prices for chlorine and for caustic soda is very difficult. Firstly, most chlorine is used captively to avoid transportation, or it is sold on the basis of long-term contracts. In relation to production, the amounts on the spot markets are small. Secondly, the production of chlorine is strictly coupled to that of the caustic

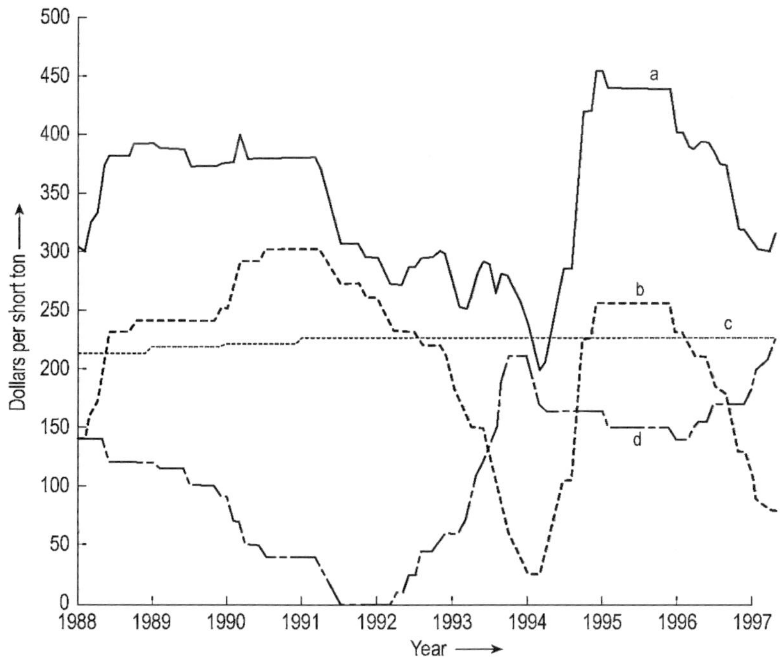

Figure 109. Production costs and product prices in the United states from 1987 to 1997 [295]
a) Product value for 1 ecu (1 t Cl_2 + 1.13 t NaOH; b) Price for caustic soda; c) Production costs for 1 ecu at 3.5 ¢/kWh electricity price; d) Price for chlorine into PVC

soda. A strong demand for chlorine creates a surplus of caustic, leading to high prices for the chlorine and a drop in prices for caustic; weak demand for chlorine reduces the available amount of caustic, leading to high caustic prices and low chlorine prices. This situation seems to change in cycles every 6 to 8 years. For example, from 1988 to 1991 the market value for chlorine fell from 150 $/t down to 0, but from March 1992 to August 1993, the price rose to 230 $/t. The price for caustic was 320 $/t in 1991; it fell to 50 $/t in March 1994 and climbed to 270 $/t in December 1994 (Fig. 109) [295].

Because of the fixed ratio of chlorine to caustic in production and the different location of uses, worldwide trade flows have been created for both products. Chlorine is traded as chlorinated derivatives, particularly as EDC, VCM, and PVC, accounting for 85 %, and chlorinated solvents. The equivalent annual amount of chlorine is approximately 4.5×10^6 t (1995 to 2000). The main exporting countries are the United States (to South East Asia, South America), Russia (to West Europe), West Europe (to South East Asia, China), Japan (to China), and Middle East Asia (to Japan, China) [296].

16. Toxicology

Chlorine gas is dangerous to health because it is a powerful oxidant. In the physiological pH range, it is converted to hypochlorous acid, a cytotoxic substance. The extent to which the cells are damaged depends on the gas concentration, the exposure time, the water content of the tissue, and the health of the person exposed to the gas. Besides getting into the eyes, larynx, and trachea, chlorine also reaches the bronchi and the bronchioles. Because of its moderate solubility in water, chlorine affects the alveoli only at high concentrations. In this respect, chlorine differs from other gases with low water solubility and high lipid solubility, such as phosgene, nitrogen monoxide, and nitrogen dioxide. Initial moderate bronchial irritation is followed by the development of a toxic pulmonary edema because of increased alveolar injury.

The olfactory threshold of chlorine gas is 0.2 – 3.5 mL/m^3. Prolonged exposure seems to raise the olfactory threshold. Concentrations of 3 – 5 mL/m^3 are tolerated for up to 30 min without any subjective feeling of malaise. At concentrations between 5 and 8 mL/m^3, mild irritation of the upper respiratory tract and the conjunctiva is observed. In addition, running of the eyes and coughing are observed at concentrations of 15 mL/m^3 and higher. Above 30 mL/m^3, the following symptoms are observed: nausea, vomiting, oppressive feeling, shortness of breath, and fits of coughing, sometimes leading to bronchial spasms. Exposure to 40 – 60 mL/m^3 leads to the development of toxic tracheobronchitis. After a latent period of several hours with fewer symptoms, pulmonary edema may occur because of alveolar membrane destruction. This is indicated by increased shortness of breath, restlessness, and cyanosis. Subsequently, a further complication can occur after several days in the form of pneumonia caused by superinfection of the injured pulmonary tissue.

A clear dose – effect relationship of chlorine gas at different concentrations in humans has not yet been published. On the basis of results obtained from animal experiments, the LC_{50} for healthy humans is assumed to be 300 – 400 mL/m^3 at a 30-min exposure [297]. No deaths have occurred in animal experiments at 30-min exposures for concentrations below 50 mL/m^3. Death following acute intoxication is caused by a fulminant pulmonary edema.

Many investigations deal with the toxicity of *low chlorine concentrations.*Investigations in humans indicate the possibility of reversible damage to the lung function parameters at concentrations starting at 1.0 mL/m^3 [298], in monkeys no effects have been observed at 0.5 mL/m^3 [299]. Long-term investigations of workers exposed to chlorine, e.g., in chlor-alkali electrolysis plants or in pulp manufacture, however, do not indicate increased rates of mortality or morbidity caused by pulmonary diseases [300] – [303].

No indications of *carcinogenicity* or *mutagenicity* of chlorine have been detected in animal experiments or encountered in industrial medicine. The MAK value is 0.5 mL/m^3 (1.5 mg/m^3); the TLV is 0.5 mL/m^3 (1.5 mg/m^3) with an STEL of 1.0 mL/m^3 (3 mg/m^3).

17. Chlorine—the Past and the Future

The first hundred years of chlorine production by chlor-alkali electrolysis may be reviewed in the light of the challenges faced by the industry. The first thirty years were mainly concerned with development of the technology. A remarkable diversity of cell components and cell arrangements were developed and tested; in mercury cells rocking cells gave way to cells with mercury pumps, with both vertical rotating and horizontal cathodes and horizontal or vertical denuders. In diaphragm cell technology both horizontal and vertical electrode arrangements were tried with diaphragms in porous concrete, asbestos paper, asbestos fibres and finally asbestos free fibres.

Materials of construction to resist wet chlorine were developed from the original ceramics, glass, plumbum, and stoneware, all of limited resistance and very difficult to seal against leaks. These materials were replaced over the last fifty years by rubber lined steel, and later by chlorinated polymers such as PVC and PVDC and then polyesters, PTFE and titanium.

Anodes were fabricated progressively in magnetite, Acheson graphite, sintered graphite and finally in coated titanium.

In the thirties a period of consolidation was started, chlorine production increased steadily, the two major products chlorine and caustic became recognised as important raw materials and the chlorine production of a country became a reliable indicator of its prosperity and of the capability of its chemical industry.

As the industry matured cooperation between producers developed in the fields of safety, technical methods of using chlorine, analytical procedures, product specifications and technical equipment.

In the sixties the public attitude towards the chemical industry in general and the chlorine industry in particular changed dramatically for the worse. Heavily publicised incidents such as Seveso and Minimata and the effects of ozone depletion and DDT plus frequent news about the suspected carcinogenic properties of many organo-chlorine compounds created a very hostile attitude in the general public and the media.

The chlorine industry made great efforts to react to the criticisms being levelled at it, however it soon became clear that the cost of implementing all the changes that were being demanded by the environmental lobby would be beyond the financial capacity of the chlorine producers. As a result the industry has concentrated on those areas that it considered to be the most important. Mercury cell operators made significant reductions in mercury control by reducing mercury losses to products, wastes, and to the atmosphere. Diaphragm cell operators made big improvements to the handling of asbestos followed by the introduction of diaphragms containing much reduced or no asbestos. The use of asbestos for diaphragms ist the only remaining use authorized by the EU. The work to produce improved operating philosophies began in the sixties and while still continuing in some countries has proved to be very effective (see Chapter 5).

JOHN RYLANDS
UNIVERSITY
LIBRARY OF
MANCHESTER

Because the chlorine industry has a tradition of mutual cooperation in matters of safety and the environment, the process of improvement was soon taken to international level. In Europe the Bureau International Technique du Chlore (BIT) was established in 1954 and in 1989 renamed as Euro Chlor. 1991 Euro Chlor was expanded to combine various chlorine-related associations for derivatives, solvents and paraffins and today is an affiliate of the European Chemical Industry Council (CEFIC). In the USA, work coordinated by the Chlorine Institute and the Chlorine Chemistry Council.

It became apparent during this long period of operational and environmental improvement that the poor image caused by Seveso and the other problems mentioned earlier was not going to be readly erased and the chlorine industry is still suffering from its earlier public relations mishaps. National and regional authorities have reacted to the media and "green" pressure by enacting a series of laws. International conventions have been implemented to protect specific areas of concern. The European Commission issued directives binding the relevant states in Europe and the United Nations has been active with an Environmental Program (UNEP) and through the United Nations Economic Commission for Europe (UN-ECE).

By the end of the 20th century much has been achieved both by regulation and by industry initiated improvements. The progress may be demonstrated by the following:

The production and use of POPs (persistent organic pollutants) such as aldrin, dieldrin, DDT, and others has been discontinued.

The production and use of certain industrial chemicals such as carbon tetrachloride, CFCs, chlorinated biphenyls and the use of elemental chlorine in the pulp and paper industry has been severely curtailed.

Emissions of mercury and asbestos have been massively reduced to an acceptable level.

The combustion of chlorinated wastes at sea ceased.

The safety of chemical plants and transportation of chlorine has been improved to a great extent.

In the new plants the membrane process has allowed the production of chlorine without the use of mercury and asbestos and with a significant reduction in electrical power consumption.

It is unlikely that in the future the chlorine industry will be looked upon favourably by the media, the environmental lobby and the general public, the best it can hope to achieve is to be tolerated because its products are essential for what are seen to be the requirements of modern life. These products are extensively used in construction, the automobile industry, pharmaceuticals, water treatment and many other areas. It is very difficult to foresee alternative materials being developed for use in these industries. The future of the chlorine industry is likely to be secure as a supplier of an essential raw material in many industries.

18. References

Note: Actual publications and informations of chlor-alkali technology and safety can be obtained from:

- The Chlorine Institute, Inc., 2001 "L" Street, N.W. Suite 506, Washington D.C., 20036, Fax 001 202 223 7225
- Euro Chlor Ave., E. van Nieuwenhuyse 4, Box 2, B-1160 Brussels, Fax 00322 6767241; e-mail: eurochlor@cefic.be

General References

[1] K. J. Vetter: *Elektrochemische Kinetik,* Springer Verlag, Berlin-Heidelberg 1961.

[2] J. S. Sconce: *Chlorine, Its Manufacture, Properties and Uses,* Reinhold Publ. Co., New York 1962.

[3] *Kirk-Othmer, Encyclopedia of Chem. Technol.,* **vol. 5,** Wiley, New York, 1993.

[4] P. Gallone, G. Milazzo: *Electrochemistry,* Elsevier, Amsterdam-London-New York 1963.

[5] C. H. Hamann, W. Vielstich: *Elektrochemie,* 3rd ed., Wiley-VCH, Weinheim 1997.

[6] *Gmelin,* system no. 6, Chlor (1927).

[7] *Winnacker – Küchler,* 3rd ed., **vol. 1,** pp. 228 – 239;
Winnacker – Küchler, 4th ed., **vol. 2,** pp. 379 – 480.

[8] R. Powell: *Chlorine and Caustic Soda Manufacture,* Noyes Data Corp., Park Ridge, N.J., 1971.

[9] R. G. Smith: *Chlorine; An Annotated Bibliography,* The Chlorine Institute, New York 1971.

[10] A. Kuhn: *Industrial Electrochemical Processes,* Elsevier, Amsterdam 1971.

[11] P. Gallone: *Trattato di ingegneria elettrochimica,* Tamburine Editore, Milano 1973.

[12] D. W. F. Hardie, W. W. Smith: *Electrolytic Manufacture of Chemicals from Salt,* The Chlorine Institute, New York 1975.

[13] A. Schmidt: *Angewandte Elektrochemie,* Verlag Chemie, Weinheim 1976.

[14] Society of Chemical Industry: *Diaphragm Cells for Chlorine Production,* London 1977.

[15] J. J. McKetta, W. A. Cunningham: *Encyclopedia of Chemical Processing and Design,* **vol. 7,** Marcel Dekker, New York-Basel 1977.

[16] R. Dandres: *Le Chlore,* 3rd ed., Institut national de recherche et de sécurité pour la prévention des accidents du travail et des maladies professionelles, Paris 1978.

[17] M. O. Coulter: *Modern Chlor-Alkali Technology,* **vol. 1,** Ellis Horwood, London 1980.

[18] C. Jackson: *Modern Chlor Alkali Technology,* **vol. 2,** Ellis Horwood, Chichester, Great Britain, 1983.

[19] K. Wall, *Modern Chlor-Alkali Technology,* **vol. 3,** Ellis Horwood, Chichester, Great Britain, 1986.

[20] N. M. Prout, *Modern Chlor-Alkali Technology,* **vol. 4,** Elsevier Appl. Science, London 1990.

[21] T. C. Wellington, *Modern Chlor-Alkali Technology,* **vol. 5,** Elsevier Appl. Science Barking, 1992.

[22] M. G. Beal in R. W. Curry (ed.): *Modern Chlor-Alkali Technology,* **vol. 6,** SCI, London 1995.

[23] S. Sealey: *Modern Chlor-Alkali Technology,* **vol. 7,** The Graham House, Cambridge 1998.

[24] The Chlorine Institute, Pamphlet 1 "Chlorine Manual", Ed. 6, 1997.

[25] Int. Soc. Electrochem.: *Extended Abstracts,* 23rd Meeting, Stockholm 1972.

[26] The Electrochemical Society: *Extended Abstracts of Industrial Electrolytic Division,* The Chlorine Institute, Spring Meeting, San Francisco,May 12nd – 17th, 1974.

[27] Int. Soc. of Electrochemistry: *Extended Abstracts,* 27th Meeting, Zürich 1976.

[28] Technischer Arbeitskreis Elektrolyse: *Hinweise für den Umgang mit Chlor,* Verlag Chemie, Weinheim 1978.

[29] The Electrochemical Society: *Extended Abstracts, Spring Meeting Seattle,* Princeton, N.J., May 1978.

[30] AIChE: *Sympos. Ser. 75, no. 185,* 1979.

[31] Oronzio de Nora Symposium: *Chlorine Technology,* Venezia, May 1979.

[32] AIChE: *Sympos. Ser. 77, no. 204,* 1981.

[33] J. S. Robinson: *Chlorine Production Processes,* Noyes Data Corp., Park Ridge, N.J., 1981.

[34] The Chlorine Institute: "Joint Chlorine Institute: BITC Meeting, London, June 1982", New York.

[35] Elektrochemische Verfahrenstechnik, *Dechema Monographie* **vol. 97,** Verlag Chemie, Weinheim 1984.

[36] Technische Elektrolysen, *Dechema Monographie* **vol. 98,** Verlag Chemie, Weinheim 1985.

Specific References

[37] C. W. Scheele, *Sv. Akad. Handl.* **35** (1774) 88.

[38] J. Eidem, L. Lunevall in [26]

[39] R. E. Means, T. R. Beck, *Chem. Eng. (N.Y.)* **91** (1984) no. 22, 46 – 51.

[40] The Chlorine Institute, Pamphlet 72: "Properties of Chlorine in SI Units", 2nd ed., July 1986.

[41] J. J. Martin, D. M. Longpre, *J. Chem. Eng. Data* **29** (1984) 466 – 473.

[42] L. Ziegler, *Chem. Ing. Techn.* **22** (1950) 229.

[43] E. M. Aleta, P. V. Roberts, *J. Chem. Eng. Data* **31** (1986) 51 – 54.

[44] I. A. A. Ketalaar, *Electrochem. Technol.* **5** (1967) 143.

[45] R. F. Schwab, N. H. Doyle, *Electrochem. Technol.* **5** (1967) 228.Y. E. Frolov, A. S. Maltsera, A. N. Baratov, A. I. Roslovskii, *Khim. Promst (Moscow)* **7** (1977) 530 –532.

[46] The Chlorine Institute, Pamphlet 121: "Explosive Properties of Gaseous Mixtures Containing Hydrogen and Chlorine", 2nd ed., June 1992.

[47] IUPAC, *White Book on Chlorine* **68** (1996) 9.

[48] V. M. Sedivy, *Industrial Minerals,* **343** (1996) April, 73.

[49] Hoechst, EP 35 695-B1, 1983.

[50] Solvay & Cie., EP 30 756-A1, 1981.

[51] A. I. Postoronko et al., SU 537 027-A, 1976. Jamestown Chem.: *Calcium Sulphate Inhibitor.*

[52] S. L. Fuks, T. M. Orchinnikova, *Zh. Prikl. Khim (Leningrad)* **56** (1983) no. 8, 1757 – 1761.

[53] K. Harada, Cleveland Meeting: *The Electrochem. Soc.,* May 1966.

[54] Olin, US 4 277 447-A, 1981 (R. F. Chambers, N. Rachima).

[55] Toa Gosei Chem. Ind., JP 57-106 520-A2, 1982 (K. Suzuki, R. Nagai, H. Goto).

[56] Hoechst, DE 3 037 818-A1, 1982 (S. Benninger).

[57] Bayer, DE 3 216 418-A1, 1983 (R. Schäfer).

[58] Bayer, DE 2 709 728-C2, 1981 (R. Schäfer).

[59] K. Maycock et al.,SCI Chlorine Symposium, London, June 1997.

[60] Hoechst, DE 2 905 125-A1, 1980 (M. Schott, J. Russow).

[61] Z. I. Lifatova et al., SU 865 800-A1, 1981.

[62] Bayer, DE 1 177 122-B, 1964 (E. Zirngiebl).

[63] Ebara Infilco, JP 54-81 174-A2, 1979 (M. Kanekawa, K. Akagi).

[64] *Processing,* June 1978.

[65] W. Glas, *Chem. Ing. Tech.* **43** (1973) 832.

[66] F. Lüns, *BBC-Nachrichten,* 1/2 (1974) 36 – 42.

[67] H. A. Horst, *Chem. Ing. Tech.* **43** (1971) 164.

[68] Siemens, DE 3 123 877-C1, 1983 (G. Achilles). Hundt & Weber Schaltgeräte, EP 11 820-A1, 1980 (E. Stratmann).

[69] F. Lappe, *Chem. Ing. Tech.* **42** (1970) 1228.

[70] Fiat Final Report 732, 797, 816, 834.

[71] *Ullmann,* 4th ed., **19,** 665.

[72] W. Kramer in [36]

[73] T. Sugino, K. Aoki, *J. Electrochem. Soc. Jpn.* **27** (1959) no. 1 – 3, E 17.

[74] G. Hauck, W. Dürr, *Chem. Ing. Tech.* **39** (1967) 720.H. J. Antweiler, J. P. Schäfer, *Chem. Ing. Tech.* **43** (1971) 180.

[75] D. S. Nielson, O. Platz, *Reliabil. Eng.* **4** (1983) 1 – 18.

[76] PPG Ind., US 4 155 819-A, 1979 (W. W. Carlin).

[77] Diamond Shamrock, US 4 073 706-A, 1978 (Z. Nagy).

[78] H. Tsukuda et al., *J. Electrochem. Soc. Jpn.* **32** (1964) 1.

[79] D. W. Wabner, P. Schmittinger, F. Hindelang, R. Huß, *Chem. Ing. Tech.* **49** (1977) 351.

[80] L. J. J. Janssen in [18]p. 271.

[81] Society of Chemical Industry, Electrochemical Technology Group: *Intern. Meeting on Electrolytic Bubbles,* 13 – 14 Sept. 1984, London.

[82] BASF, US 3 502 561, 1970 (W. Rasche, E. Wygasch, G. Csizi).

[83] A. Zimmer, L. Franke, *Chem. Tech. (Leipzig)* **36** (1984) no. 2, 64 – 69.

[84] H. Hund, *Chem. Ing. Tech.* **39** (1967) 702.

[85] W. Vielstich, *Chem. Ing. Tech.* **34**(1962) 346. Bayer, DE 1 792 588-C3, 1977 (E. Zirngiebl).Murgatroyd's Salt & Chem., DE 1 915 765, 1969 (M. J. Lockett).

[86] M. M. Jaksic, I. M. Csonka, *Electrochem. Technol.* **4** (1966) 49.

[87] G. Barthel, *Chem. Anlagen + Verfahren* **8** (1973) 59 – 60.

[88] P. Theissing et al., *Chem. Ing. Tech.* **51** (1979) 237.F. Hine: *Topics in Pure and Applied Electrochemistry,* SAEST, Karaikudi, India.

[89] Uhde: *Alkalichloridelektrolyse nach dem Quecksilberverfahren.*

[90] Olin, US 4 004 989-A, 1977 (R. W. Ralston).

[91] G. Hauck, *Chem. Ing. Tech.* **34** (1962) 369.

[92] Euro Chlor: "Control of Worker Exposure to Mercury in the Chlor-Alkali Industry", 4th ed., October 1997.

[93] The Chlorine Institute: Pamphlet 125, Guidelines: "Medical Surveillance and Hygiene Monitoring Practices for Control of Workers Exposure to Mercury in the Chlor-Alkali Industry", 3rd ed., December 1994.

[94] Environmental Protection Agency: *Control Techniques for Mercury Emissions from Extraction and Chlor-Alkali Plants,* Publ. No. AP-118, Washington, DC.

[95] A. Gellera, L. Cavalli, G. Nucci, *Analyst (London)* **109** (1984) 1537 – 1539.

[96] H. Klotz, *Chem. Ing. Tech.* **52** (1980) 444.

[97] Akzo: *Anode Current Control by Computer ACCS-System,* 1972.

[98] A. I. Vasilev et al., SU 1 024 528-A1, 1983.

[99] Bayer, DE 1 767 840-B2, 1975 (R. Schlee, W.Büsing).Hoechst, DE 3 244 033-A1, 1984 (K. Lehmann, H. Valentin, K. Gnann).

[100] Montedison, EP 93 452-A2, 1983 (F. Lo Vullo, E. Malvezzi, P. Balboni).

[101] Solvay & Cie., EP 85 999-A1, 1983 (J. P. Detournay, J. Defourny).

[102] Heraeus Elektroden, EP 68 076-A2, 1983 (P. Fabian).

[103] H. Becker, *Dechema Monographie,* **vol. 61,** Verlag Chemie, Weinheim 1968, p. 59.

[104] BITC Working Group Mercury Analysis, *Anal. Chim. Acta* **72** (1974) 37 – 48; **84** (1976) 231 – 257; **87** (1976) 273 – 281.

[105] AKZO: *Das IMAC-TMR-Verfahren.*

[106] EEC Directive 82/176, 22. 3. 1982.

[107] Euro Chlor Publ. Env. Prot. 11 "Mercury Housekeeping", 4th ed., September 1998.

[108] Euro Chlor: "Good Mercury Housekeeping" Third Techn. Seminar, Proceedings, December 1993.

[109] Parcom Decision 90/3, Reducing Atmospheric Emissions from Existing Chlor-Alkali Plants, June 14, 1990.

[110] Euro Chlor, "Chlorine Industry Review 1996 – 1997", p. 10.

[111] A. A, Lindley, An Economic and Environmental Analysis of the Chlor-Alkali Production Process, prepared for the European Commision (DG III, Chemicals), Final Report, June 30, 1997.

[112] Euro Chlor, GEST 97/228, "The Euro Chlor Program", Fourth Techn. Seminar, Proceedings, Feb. 1997.

[113] ELTECH Systems Corporation Technical Service Department: *Diaphragm Electrolytic Chlorine Cell, Technical Data Manual,* ELTECH Systems Corporation, Chardon, Ohio 1982.

[114] ELTECH Systems Corporation Technical Service Department: *Diaphragm Chlorine Cell, Operating Manual,* ELTECH Systems Corporation, Chardon, Ohio 1983.

[115] ELTECH Systems Corporation, US 4 410 411, 1983 (R. W. Fenn III, E. J. Pless, R. L. Harris, K. J.O'Leary).

[116] ELTECH Systems Corporation, US 3 674 676, 1972 (Edward I. Fogelman).

[117] L. C. Curlin, T. F. Florkiewicz, R. C. Matousek: "Polyramix, A Depositable Replacement for Asbestos Diaphragms," London International Chlorine Symposium, 1988.

[118] R. N. Beaver: *The Dow Diaphragm Cell,* Dow Chemical, Freeport, Texas 1985.

[119] Dow Chemical Company, US 4 497 112.

[120] Oronzio De Nora Impianti Elettrochimici S.p.A.: *Design Features of Glanor Diaphragm Electrolyzer,* Oronzio DeNora Impianti Elettrochimici S.p.A., Milano.

[121] PPG Industries: *PPG's Bipolar Diaphragm Electrolyzers,* PPG Industries, Inc., Pittsburgh, P.

[122] PPG Industries: *Licensing From PPG Chemicals,* PPG Industries, Inc., Pittsburgh, Pa.

[123] Occidental Chemical Corporation: *Hooker Chlor-Alkali Systems, A Record of Achievement in Diaphragm Cell Technology,* Occidental Chemical Corporation, Niagara Falls, N.Y.

[124] Uhde GmbH: *Alkaline Chloride Elektrolysis by the Diaphragm Process; System Hooker,* Uhde GmbH, Dortmund.

[125] ELTECH Systems Corporation: *ELTECH Modified Diaphragm Cells,* ELTECH Systems Corporation, Chardon, Ohio 1985.

[126] ELTECH Systems Corporation, US 3 591 483, US 3 707 454, 1971 (Richard E. Loftfield, Henry W. Laub).

[127] ELTECH Systems Corporation, US 3 928 166, 1975 (Kevin J. O'Leary, Charles P. Tomba, Robert W. Fenn III).

[128] ELTECH Systems Corporation: *Caustic Purification System 1985.*

[129] G. A. Carlson, E. E. Estep, "Porous Cathode Cell for Metals Removal from Aqueous Solutions," paper presented at the Electrochemical Society Meeting, May, 1972.

[130] W. G. Grot, *Chem. Ing. Tech.* **44** (1972) 167; Diamond Shamrock, US 4 025 405, 1977 (R. L. Dotson, K. J. O'Leary).

[131] Asahi Glass, US 4 065 366, 1977 (Y. Oda, M. Suhara, E. Endoh).

[132] H. Ukihashi, M. Yamabe, *ACS Polymer Division Workshop on Perfluorinated Ionomer Membranes,* Lake Buena Vista, Florida, Feb. 23 – 26, 1982.

[133] *World Chlor-alkali Analysis,* **vol. 1,** CMAI, Texas, USA May 1997.

[134] T. A. Davis, J. D. Genders, D. Pletcher, *Ion permeable membrane,* Alresford 1997.

[135] Asahi Chemical Industry, US 4 178 218, 1979 (M. Seko).

[136] K. Schneiders, J. Schulze: Improvement in the Hoechst-Uhde Membrane Cell, *Modern Chlor-Alkali-Technology,* **vol. 6,** 1995.

[137] Akin, *"Optimierung einer Chloralkalielectrolyse-Membranzellen-Kaskade unter Berücksichtigung ihres Energieverbrauchs",* Dissertation, Dortmund 1984.

[138] Tokuyama Soda, US 4 190 514, 1980 (S. Matsuura, T. Oku, M. Kuramatani, N. Kramoto, Y. Ozaki).

[139] PPG, US 4 323 595, 1982 (C. N. Welth, J. O. Snodgrass).

[140] H. S. Burney: Chlor-Alkali Process, edited by Ralph. E. White et al., *Modern Aspects of Electrochemistry, Number 24,* Plenum Press, New York, 1993.

[141] Du Pont: *Recommended Brine Impurity Limits for Chloralkali Plants,* Aug. 1991.

[142] Bayer AG: *The Importance of Electrolyte Quality to Membrane Chloralkali Cells,* Uhde Chlorine Symposium 1995 (F. R. Minz).

[143] Du Pont: *Effect of Impurities on Membrane for Chloralkali Production,* Wilmington, Delaware, Aug. 1989 (James T. Keating).

[144] Rolf Rodermund: *Chlorat-Reduktion im Solekreislauf einer Chlor-Alkali-Electrolyse,* Dissertation, Dortmund 1983.

[145] Asahi Glass, US 4 255 523, 1981 (H. Ukihashi, T. Asawa, T. Gunjima).

[146] Du Pont, US 4 021 327, 1977 (W. G. Grot).

[147] Asahi Glass, JP 59/40 231, 1984 (Y. Oda, T. Morimoto, K. Suzuki).

[148] The Dow Chemical Company: *Membrane Chlor-Alkali Process,* Texas Applied Science and Technology Laboratories, Freeport, Texas, 1993 (H. S. Burney).

[149] D. Bergner, *Chem.-Ztg.* **107** (1983) 281.

[150] H. B. Beer, GB 1 147 442, 1965.H. B. Beer, GB 1 195 871, 1967.

[151] V. de Nora, J.-W. Kühn von Burgsdorff, *Chem. Ing. Tech.* **47** (1975) 125 – 128. Electronor Corporation, US 3 616 445, 1967 (G. Bianchi, V. de Nora, P. Gallone, A. Nidola).

[152] S. Trasatti, G. Lodi in S. Trasatti (ed.): *Electrodes of Conductive Metallic Oxides,* part A, Elsevier Scientific Publishing Company, Amsterdam-Oxford-New York 1980, pp. 301 – 358.

[153] S. Trasatti, W. E. O'Grady in H. Gerischer, C. W. Tobias (eds.): *Advances in Electrochemistry and Electrochemical Eng.,* **vol. 12,** J. Wiley & Sons, New York-Chichester 1981, pp. 177 – 261.

[154] GB 1 402 414, 1971(N. W. J. Pumphrey, B. Hesketh).GB 1 484 015, 1973(B. Hesketh, C. Pownall, N. W. J. Pumphrey).

[155] C. Conradty, DE-OS 1 813 944, 1968 (G. Thiele, D. Zöllner, K. Koziol); C. Conradty, DE 2 255 690, 1972 (K. Koziol, K.-H. Sieberer, H.-C. Rathjen).

[156] C. Modes, W. C. Heraeus: abstract and poster during the *13th Int. Congress and General Assembly,* Int. Union of Crystallography, Hamburg, August 1984.

[157] Yu. E. Roginskaya, O. V. Morozova, *Electrochimica Acta* **40** (1995) 817 – 822.

[158] D. V. Kokoulina, T. V. Ivanova, Y. I. Krasavitskaya, Z. I. Kudryavtseva et al., *Elektrokhimiya* **13** (1977) no. 10, 1511 – 1515.

[159] J. E. Currey in *Encycl. of Chem. Proc. and Design,* **vol. 7,** Marcel Dekker, New York 1978, pp. 305 –450.

[160] D. Bergner, S. Kotowski, *J. Appl. Electrochem.* **13** (1983) 341 – 350.

[161] Y. M. Kolotyrkin, *Denki Kagaku* **47** (1979) no. 7, 390 – 400.

[162] D. W. Wabner, P. Schmittinger, F. Hindelang, R. Huß, *Chem. Ing. Tech.* **49** (1977) 351.

[163] B. Busse, R. Scannell, *Electrochemische Verfahrenstechnik-Energietechnik, Stoffgewinnung und Bioelektrochemie,* Mono. GdCh. Tagung Monheim 1996.

[164] F. Beck, *Electrochimica Acta* **34** (1989) 811 – 822.

[165] Oronzio De Nora Impianti Elettrochimici S.p.A., GB 2 051 131, 1980 (A. Pellegri).

[166] A. Nidola in S. Trasatti (ed.): *Electrode of Conductive Metallic Oxides,* part B, Amsterdam-Oxford-New York 1981, pp. 627 – 659.

[167] C. Traini, *Proc. of O. De Nora Symposium,* 15 – 18 May 1979.

[168] Diamond Shamrock Corp., US 3 674 676, 1970 (E. I. Fogelman).

[169] V. de Nora, *Chem. Ing. Tech.* **47** (1975) no. 4, 141.

[170] Heraeus Elektrochemie, EP 0 687 312, 1994 (B. Busse, R. Scannell).

[171] S. Kotowski, B. Busse, paper presented at the Annual DECHEMA Meeting 1984, *Anodenbeschichtungen in Membranzellen zur Chlor- und Laugeproduktion.*

[172] Uhde GmbH, DE 3 219 704, 1982 (H. Schmitt, H. Schurig, W. Strewe).

[173] N. P. Fedot'ev, N. V. Berezina, E. G. Kruglova, *Zh. Prikl. Khim. (Leningrad)* **21** (1948) 317 – 328.

[174] Toa Gosei Chemical Industry Co. Ltd., JP 31-6611, 1956 (K. Sasaki).

[175] E. I. DuPont de Nemours & Co., US 4 116 804, 1978 (C. R. S. Needes).

[176] Diamond Shamrock Corp., US 4 024 044, 1977 (J. R. Brannan, I. Malkin).

[177] F. Hine, M. Yasuda, M. Watanabe: "Studies of the Nickel – Sulfur Electrodeposited Cathode," *Denki Kagaku* **47** (1979) no. 7, 401 – 408.

[178] Asahi Chemical, EP-A 0 031 948, 1981 (M. Yoshida, H. Shiroki).

[179] Chlorine Engineers Corp. Ltd., US 4 465 580, 1984 (K. Kasuya).

[180] Tokuyama Soda, US 4 190 516, 1980 (Y. Kajimaya, T. Kojima, Y. Murakami, S. Matsura).

[181] Dow Chemical, EP-A 0 129 734, 1985 (N. R. Beaver, L. E. Alexander, C. E. Byrd).

[182] ICI, EP-A 0 129 374, 1984 (J. F. Cairns, D. A. Denton, P. A. Izard).

[183] Diamond Shamrock Corp., US 4 104 133, 1978 (J. R. Brannan, I. Malkin, C. M. Brown).

[184] Kanegafuchi Kagaku Kogyo Kabushiki Kaisha,EP-A 0 132 816, 1985 (Y. Samejima, M. Shiga, T. Kano, T. Kishi).

[185] Olin Corp., US 4 169 775, 1979 (H. C. Kuo).

[186] Harrimann Chemsult: Chlor-Alkali Prospects for Chlorine, Caustic Soda and Soda Ash in the 1990's, London 1991.

[187] Th. F. Florkiewicz, *The Future of Diaphragm Cell Technology,* Chlorine Institute Spring Meeting, St. Louis, Missouri, March 1998.

[188] M. Esayian, J. H. Austin: "Membrane Technology for Existing Chloralkali Plants," E. I. du Pont de Nemours Co., Wilmington, Del., 1984.

[189] VCI Verband der Chemischen Industrie (ed.): *Zukunft der Alkalichlorid-Electrolyse-Anlagen nach dem Amalgamverfahren,* Frankfurt a. M. 1998.

[190] M. R. Marcan: "Mercury/Membrane Cells — when to Change", *Modern Chlor-Alkali Technology,* **vol. 5,** Elsevier Publ., London, 1992, p. 177.

[191] R. W. Curry: "Issues involved in membrane conversion of a chlor-alkali business", *Modern Chlor-Alkali Technology,* **vol. 5,** Elsevier Publ., London 1992, p. 185.

[192] S. Higuchi, *Soda to Enso* **35** (1984) 383.

[193] H. Isfort, W. Stockmans in [36]

[194] K. Kerger, *Chem. Ing. Tech.* **43** (1971) 167.

[195] Uhde: *Chlor und Wasserstoff aus Salzsäure durch Elektrolyse.*

[196] General Electric, DE 2 856 882-A1, 1979 (R. M. Dempsey, A. B. Conti).

[197] H. Klotz, P. Orth in [36]

[198] The M. W. Kellogg Co.: *Hydrogen Chloride to Chlorine, The Kel-Chlor Process,* presentation at the 20th Chlorine Plant Managers Seminar, New Orleans, Feb. 9, 1977.

[199] Krupp Uhde company brochure "Chlorine and Hydrogen from Acid by Electrolysis," 1993.

[200] *Chemical Week* 1996, 14. Aug., 26.
[201] The M. W. Kellogg Co., *Hydrocarbon Process.* 60 (1981) no. 11, 143.
[202] Shell, BE 599 241-A, 1961 (W. S. Engel, F. Wattimena).
[203] Mitsui Chemicals Inc.: "The Mitsui MT-Chlor Process, Catalytic Oxidation of Hydrogen Chloride," company brochure.
[204] M. Mortensen et al., *Chem. Engng. Scie.* **51** (1996) 2031.
[205] Institut Français du Pétrole, FR 1 294 706-A, 1962 (P. Bedagne et al.).
[206] C. P. van Dijk, *Chem. Econ. Eng. Rev.* **4** (1972) no. 12, 42.
[207] H. Schmidt, F. Holzinger, *Chem. Ing. Tech.* **35** (1963) 37.
[208] Monsanto Enviro-Chem. Systems: *Fiber Bed Mist Eliminator,* 1979.
[209] Dow Chemical, US 3 568 409-A, 1971 (B. Mac Ferguson, J. F. Gilbert, D. N. Glew).
[210] The Chlorine Institute, Pamphet 134: "The Drying and Liquefication of Chlorine and the Phase Diagram Cl_2-H_2O", April 1967.
[211] Koninkl. Nederlandsche Zoutindustrie, US 3 534 562-A, 1970 (D. Meyer, T. Thijssen).
[212] ICI, DE 1 946 096-C3, 1973 (K. J. Howliston).
[213] G. Alberti, *DECHEMA Monogr.* **66** (1971) 1193–1211. R. Gabbioneta: *Garo.* Nash International: *HC-7 Chlorine Compressor.*
[214] Sulzer Burckhardt, Prospectus No. 1393 a.
[215] Sundstrand: *Sundyne Zentrifugalverdichter,* 1980.
[216] Euro Chlor 79/79 "Chlorine Transfer Compressors", 3rd ed. February 1999.
[217] Euro Chlor 83/119 "Canned Pumps for Use with Liquid Chlorine", 3rd ed. September 1987.
[218] H. Abel, F. Özvegyi, *Kälte-Klima-Rundsch.* **7** (1969) 100.
[219] *Inf. Chim.* **98** (1971) 55.
[220] H. Hagemann, *Chem. Ing. Tech.* **39** (1967) 744.
[221] R. F. Schwab, W. H. Doyle, *Electrochem. Technol.* **5** (1967) 228.
[222] B. Liggenstorfer: "Kälteanlage zum Verflüssigen von Chlor," *Tech. Rundsch. Sulzer* **63** (1981) no. 2, 53–56.
[223] AKZO: *Chlorine Liquefaction by Absorption and Distillation,* 1974.
[224] Euro Chlor Position Paper PP VII: "Use of Carbontetrachloride in Chlorine Production", Feb. 1995.
[225] T. A. Liederbach, *Chem. Eng. Prog.* **70** (1974) 64.
[226] Euro Chlor: "Publications, Catalog", Brussels, Febr. 1997.
[227] The Chlorine Institute: "Technical and Safety Information, Catalog 1997/98".
[228] R. Dandres: *Le Chlore,* 3rd ed., L'Institut National de Recherche et de Sécurite pour la Prevention des Accidents du Travail et de Maladies Professionelles, Paris 1978. H. Dorias: *Gefährliche Güter,* Springer Verlag, Berlin-Heidelberg-New York-Tokyo 1984.
[229] G. Payne: "Safe Handling and Storage of Liquefied Gases, in Particular Chlorine and Chlorofluorohydrocarbon Refrigerants," *Trans. Inst. Chem. Eng.* **42** (1964) no. 4, 92–99.
[230] D. P. Meinhardt, "Chlorine," *Chem. Eng.* (1981) Oct. 5, 125–133.
[231] N. C. Harris, J. P. Shaw, *European Chlorine Storage Practice,* paper presented at the Chlorine Institute's 23rd Plant Manager Seminar, 1980. R. Papp, *Chlorine '74,* Chemistry and Industry, 15 March (1975) 243–246.
[232] Euro Chlor Techn. Publ. 75/46, "Pneumatically Operated Valves for Use on Rail and Road Tankers, and ISO-Containers for Liquid Chlorine", 8th ed., November 1996.
[233] The Chlorine Institute, Pamphlet 93, "Pneumatically Operated Valves for Use on Tank Cars", Ed. 1, September 1996.
[234] *Winnacker-Küchler:* 4th ed., pp. 428–429.

[235] Euro Chlor Techn. Publ. 72/10, "Pressure Storage of Liquid Chlorine", 8th ed., June 1993.

[236] The Chlorine Institute, Pamphlet 5, "Non-refrigerated Liquid Chlorine Storage", Ed. 5, October 1992.

[237] Euro Chlor Techn. Publ. 73/17, "Low Pressure Storage of Liquid Chlorine", 5th ed., February 1996.

[238] The Chlorine Institute, Pamphlet 78, "Refrigerated Liquid Chlorine Storage", Ed. 2, June 1994.

[239] ISO 3511/1.

[240] A. F. Timofeer et al.: "Steps to Prevent Failures of Chlorine Containers," *Khim.Promst* (Moscow) 10 (1978) no. 3, 202–205; Sov. Chem. Ind. (Engl. Transl.) 10 (1978) no. 3, 222–225.

[241] Euro Chlor Techn. Publ. 79/81, "Dry Liquid Chlorine Piping Systems Located inside Producers' or Consumers' Plants", 7th ed., September 1986.

[242] The Chlorine Institute, Pamphlet 60, "Chlorine Pipelines", Ed. 4, May 1997.

[243] The Chlorine Institute, Pamphlet 9, "Chlorine Vaporizing Equipment", Ed. 5, April 1997.

[244] Euro Chlor, Techn. Publ. 75/47, "Chlorine Vaporizers", 8th ed., January 1980.

[245] Armstrong Chemtec, Form 54–2/97, *Vaporizers.*

[246] Euro Chlor Techn. Publ. 76/55, "Acceptable Levels of Nitrogen Trichloride in Liquid Chlorine", 9th ed., September 1990.

[247] The Chlorine Institute, Pamphlet 21, "Nitrogen Trichloride–A Collection of Reports and Papers", Ed. 4, 1997.

[248] The Chlorine Institute, Pamphlet 89, "Chlorine Scrubbing Systems", Ed. 1, December 1991.

[249] Euro Chlor Techn. Publ. 76/52, "Equipment for the Treatment of Gaseous Effluents Containing Chorine", 9th ed., October 1986.

[250] M. E. Fakley et al., The ICI HYDECAT-Process in [21] p. 23.

[251] Euro Chlor Techn. Publ. 79/82 "Choice of Materials of Construction for Use in Contact with Chlorine", 7th ed., October 1995.

[252] W. A. Statesir: "Explosive Reactivity of Organics and Chlorine," *Chem. Eng. Prog.* **69** (1973) no. 4, 52–54.

[253] Euro Chlor Third Technical Seminar Proceedings, Techn. Publ. 93/192 "How to Use Steel and Titanium Safety", December 1993.

[254] The Chlorine Institute, Pamphlet 95, "Gaskets for Chlorine Service", Ed. 1, October 1993.

[255] T. B. Meslin: "Assessment and Management of Risk in the Transport of Dangerous Materials: The Case of Chlorine Transport in France," *Risk Anal.* **1** (1981) no. 2, 137–142.G. B. Frame: "Determination of Risk from a Chlorine Spill or Major Release," *Chem. Can.* **32** (1980) no. 8, 27–29.

[256] C. L. Melancon, *Dow Chemicals Emergency Response System,* paper presented at The Chlorine Institute's 22nd Plant Manager's Seminar 1979.R. Sklarew et al., *Emergency System for Toxic Chemical Releases,* Pollution Engineering, July 1982.

[257] J. M. Buchlin, *Aerodynamic Behaviour of Liquid Spray-Design Method of Water Spray Curtain,* von Karman Institute, Rhode Saint Genese, Belgium 1980, Test Report No. 171.

[258] The Chlorine Institute, Pamphlet 63, "First Aid and Management of Chlorine Exposures", Ed. 5, May 1993.

[259] The Chlorine Institute, Pamphlet 64, "Emergency Response Plans for Chlorine Facilities", Ed. 4, November 1995.

[260] The Chlorine Institute, Pamphlet 65, "Personal Protective Equipment for Chlorine and Sodium Hydoxide", Ed. 3, November 1995.

[261] The Chlorine Institute, Pamphlet 85, "Recommendations for Prevention of Personnel Injuries for Chlorine Producer and User Facilities", Ed. 3, November 1994.

[262] The Chlorine Institute, Pamphlet 86, "Recommendations for Chlor-Alkali Manufacturing Facilities for the Prevention of Chlorine Releases", Ed. 3, November 1994.
[263] The Chlorine Institute, Pamphlet 74, "Estimating the Area Affected by a Chlorine Release", Ed. 3, in revision.
[264] The Chlorine Institute, Instruction Booklets, "Chlorine Institute Emergency Kit "A" for 100-lb. and 150-lb. Chlorine Cylinders, Kit "B" for Chlorine Ton Containers, Kit "C" for Chlorine Tank Cars and Tank Trucks", Ed. 8, 1996.
[265] Euro Chlor Chlorine Safety Manual 90/159, "Aspects of Transport Safety", November 14–15, 1990.
[266] Euro Chlor Techn. Publ. 92/176, "Chlorine Emergency Equipment", 1st ed. September 1993.
[267] Euro Chlor Techn. Publ. 93/191, 93/193, 93/194, "Safe Design", "Safe Constructions", "Safe Operations in Euro Chlor Third Technical Seminar Proceedings", December 2–3, 1993.
[268] The Chlorine Institute, Chlorep: How to Get Help in Chlorine Emergencies.
[269] British Standard Institution: *Specification for Liquid Chlorine,* BS 3947, 1965.
[270] ISO 1552: Liquid Chlorine for Industrial Use — Method of Sampling.
[271] The Chlorine Institute, Pamphlet 77, "Sampling Liquid Chlorine", Ed. 3, April 1997.
[272] ISO 2121: Liquid Chlorine for Industrial Use — Determination of the Water Content — Gravimetric Method.
[273] ASTM–E410 *Standard Method of Test for Moisture and Residue in Liquid Chlorine,* 1986.
[274] ISO 2202: Liquid Chlorine for Industrial Use — Determination of the Water Content — using an Electrolytic Analyzer.
[275] ISO 2120: Liquid Chlorine for Industrial Use — Determination of the Content of Chlorine by Volume in the Vaporized Product.
[276] ASTM–E412 *Standard Method of Assay of Liquid Chlorine,* 1986.
[277] DIN 38 408 Teil 4:Gasförmige Bestandteile (Gruppe G) Bestimmung von freiem und Gesamtchlor (G 4).
[278] *Anal. Chim. Acta,* **156** (1984) 221–233.
[279] *Anal. Chim. Acta,* **87** (1976) 273–281.
[280] ASTM–E506 *Standard Test Method for Mercury in Liquid Chlorine,* 1984.
[281] *Deutsche Einheitsverfahren zur Wasseruntersuchung,* Verlag Chemie, Weinheim 1971.
[282] M. L. Langhorst, *Am. Ind. Hyg. Assoc. J.* **43** (1982) 347–360.The Chlorine Institute: *Atmospheric Monitoring Equipment for Chlorine,* 2nd ed., March 1982, CI Member Information Report 138.W. König et al., *Chem. Ind.* (Feb. 1984) 69.
[283] World Health Organisation, *"Recommendations for Drinking Water Quality",* 2nd ed., **Vol.1,** December 1993.W. König et al., *Chem. Ind.* **36** (Feb. 1984) 69.
[284] Mary. C. Blackburn, Chemical Market Associates, Inc.: "Chlorine and Caustic — which Way is up", Krupp Uhde Symposium, Dortmund, May 6–8, 1998.
[285] W. König, K. Krinke, H. Kron et al., *Chem. Ind.* **36** (Feb. 1984) 69.
[286] Euro Chlor, "Chlorine Industry Review 1994–1995".
[287] *Chem. Engng. News,* 1994, Nov. 21, 17.
[288] Euro Chlor, Proceedings of the Technical Seminar Leipzig "Responsible Care in Practise", Germany 1997.
[289] Euro Chlor: "Chlorine flow in Europe 1995," Jan. 1997.
[290] Euro Chlor: "Chlorine Industry Review 1994–1995".
[291] Harriman Chemsult, Chlor-Alkali Prospects for Chlorine, Caustic Soda and Soda Ash in the 1990's, Autumn 1991, London, England.
[292] B. Hileman et al., *Chem. & Eng. News,* 21st November 1994, p. 12.

[293] The Chlorine Institute, Pamphlet 10, "North American Chlor-Alkali Industry Plants and Production Data Book", April 1997. The Chlorine Institute, Pamphlet 16, "Chlor-Alkali Producers Outside North America" July 1997.

[294] SRI International – Chemical Economics Handbook, CEH Marketing Report, "Chlorine/Sodium Hydroxide", 1996, USA, Zurich, Tokyo.

[295] ANORGANICA Ltd, Monthly Newsletter "Soda – Chlorine" Issue Number 197, 29th May 1997.

[296] Euro Chlor, GEST 97/230, D. Hutchinson: "Derzeitige und zukünftige Tendenzen bei der Herstellung von Chlor, Natronlauge und Chlorderivaten", 4th Technical Seminar, Leipzig, 1997.

[297] P. Davies, I. Hymes: "Chlorine toxicity criteria for hazard assessment," *Chem. Eng.*, June 1985, 30 – 33.

[298] H. H. Rotman, M. J. Fliegelman, T. Moore, R. G. Smith, D. M. Anglen, C. J. Kowalski, J. G. Weg: "Effects of low concentrations of chlorine on pulmonary function in humans," *J. Appl. Physiol.: Respir. Environ. Excercise Physiol.* **54** (1983) no. 4, 1120 – 1124.

[299] "Chronic Inhalation Toxicity Study on Chlorine in Non-Human Primates," *CITT Activities Chemical Insitute of Toxicology* **4** (1984) no. 8, 1 – 3.

[300] B. Grenquist-Norden: "Respiratory effects of industrial chlorine and chlorine dioxide exposure," Institute of Occupational Health, University of Helsinki, Finland, 1983.

[301] F. Schuckmann (Occupational Health Department of Hoechst AG): "Pulmonary Function Study on Workers with Long-Term Exposure to Chlorine," in: *Medichem proceedings, XI. International Congress, Calgary, Alberta, Canada,* 26. – 29. 9. 1983, pp. 475 –483.

[302] B. G. Ferris, Jr., S. Puleo, H. Y. Chen: "Mortality and morbidity in a pulp and a paper mill in the United States: ten-year follow-up," *Br.J. Ind. Med.,* **36** (1979) no. 2, 127 – 134.

[303] L. R. S. Patil, R. G. Smith, A. J. Vorwald, T. F. Mooney: "The health of diaphragm cell workers exposed to chlorine," *Am. Ind. Hyg. Assoc. J.* **31** (1970) 678 – 686.

[304] J. Starnick: "Chlor har Zukunft – Die Zukunft braucht Chlor", VCI-Symposium, Berlin, February 1999.

[305] T. E. Graedel, W. C. Keene: "IUPAC White Book on Chlorine", *Pure & Appl. Chem.* **68** (1996) no. 9, 1689.

[306] Fonds der Chemischen Industrie "Die Chemie des Chlors und seiner Verbindungen" Brochure no. 24 (1992) 7.

[307] G. W. Gribble: "IUPAC Handbook on Chlorine", *Pure & Appl. Chem.* **68** (1996) no. 9, 1699.

[308] F. Behr: "Fluorine Compounds, Organic" in: *Ullmann*, 5th ed., **11,** 373.

[309] H. Miyake, Asahi Glass Co. Ltd: The design and development of Flemion membranes, *Modern Chlor-Alkali Technology,* **vol. 5,** Elsevier Appl. Science, Barking, 1992, pp. 59 – 68.

[310] Fonds der Chemischen Industrie "Die Chemie des Chlors und seiner Verbindungen", Frankfurt/Main 1992, p. 14.

[311] D. Bergner, *Uhde Annual Meeting,* Dortmund, 1992, p. 170.

[312] F. Minz: "Sodium Hydroxide" in: *Ullmann,* 5th ed., **24,** 345 – 354, and 6th ed., Electronic Release 1999.

[313] M. G. Beal: *Modern Chlor-Alkali Technology,* **vol. 7,** The Royal Society of Chemistry, Cambridge 1998, pp. 1 – 2.

[314] J. K. Nelson in B. J. Moniz, W. J. Pollock (eds): *Process Industries Corrosion,* NACE, Houston 1986.

[315] The Chlorine Institute, Pamphlet 94, "Sodium Hydroxide Solution and Potassium Hydroxide Solution (caustic): Storage Equipment and Piping System", ed. 1, Jan. 1995.

[316] The Chlorine Institute, Pamphlet 87: "Caustic Tank Cars: Loading and Unloading" ed. 1, April 1993; Pamphlet 88: "caustic Tank Motor Vehicle Loading and Unloading", ed. 1, Dec. 1994.

[317] The Chlorine Institute, Pamphlet 80, "Guidelines for Handling Caustic Soda and Caustic Potash Barges", ed. 2, Nov. 1995.

[318] M. Blackburn, CMAI: "Chlorine and Caustic, which Way is up", Proceedings of the Krupp Uhde Symposium, Dortmund, May 1998.

[318a] P. Schmittinger, "Potassium Compounds" in Ullmann, 5th ed., 22, 94.

[319] D. Hutchinson, Euro Chlor Proceedings of the 4th Technical Seminar Leipzig, February 1997, pp. 38–44.

[320] H. Galal-Gorchev, *Pure & Appl. Chem.* **68** (1996) no. 9, 1731.

[321] H. Halal-Gorchev, Euro Chlor Proceedings of the 3rd Global Chlor-Alkali Symposium, Monte Carlo, 1992, p. 148.

[322] R. H. Moser: *The Future of Chlorine in Water Purification,* Paper for the 2nd World Chlor-Alkali Symposium, Washington D.C., September 1990.

[323] W. Roeske: "Die Desinfektion von Trinkwasser mit Chlor und Chlordioxid", *Wasser, Luft und Boden* **7–8,** 1998.

[324] WHO: *Recommendations for Drinking Water Quality,* 2nd ed., **vol. 1,** December 1993.

[325] M. Monzain, Euro Chlor Proceedings of the 4th Technical Seminar Leipzig, February 1997, p. 172.

[326] The Chlorine Institute, Pamphlet 97, "Safety Guidelines for Swimming Pool Applicators", ed. 1, October 1995.

[327] B. Cyne et al.: *Modern Chlor-Alkali Technology,* **vol. 7,** The Royal Society of Chemistry, Cambridge, 1998, p. 197.

[328] R. Patt, O. Kordaschia: "Paper and Pulp" in: *Ullmann,* 5th ed., **18,** 545–593, and 6th ed., Electronic Release 1999.

[329] C. K. Brennan: *Modern Chlor-Alkali Technology,* **vol. 3,** Ellis Horwood Ltd, Chichester, 1986, p. 29.

[330] Euro Chlor "Chlorine Industry Review 1997–1998".

[331] G. Riess: "Phosphorus-Halogen Compounds" in: *Ullmann,* 5th ed., **19,** 532, and 6th ed., Electronic Release 1999.

[332] J. Svara, N. Weferling: "Phosphorus Compounds, Organic" in: *Ullmann,* 5th ed., **19,** 545.

[333] H. Lauss, W. Steffens: "Sulfur Halides" in: *Ullmann,* 5th ed., **25,** 623–634.

[334] Y. Ura, G. Sakata: "Chloroamines" in: *Ullmann,* 5th ed., **6,** 553–558.

[335] N. Kriebitzsch, H. Klenk: "Cyanuric Acid and Cyanuric Chloride" in: *Ullmann,* 5th ed., **8,** 191–200.

[336] J.-P. Schirmann: "Hydrazine" in: *Ullmann,* 5th ed., **13,** 177–191.

[337] S. Austin, A. Glowacki: "Hydrochloric Acid" in: *Ullmann,* **13,** 283–296.

[338] R. Nolte, *Die europäische Chlorbilanz,* Euro Chlor Proceedings of the 4th Technical Seminar Leipzig, 1997, p. 70.

[339] H. Vogt et al.: "Chlorine Oxides and Chlorine Oxygen Acids" in: *Ullmann,* 5th ed., **6,** 483–525.

[340] Potasse et Produits Chimiques, FR 1 019 027, 1953.

[341] H. Sibum et al.: "Titanium, Titanium Alloys and Titanium Compounds" in: *Ullmann,* 6th ed., Electronic Release 1999.

[342] H. Sibum: "Titanium and Titanium Alloys" in: *Ullmann,* 5th ed., **27,** 100.

[343] R. Nielsen: "Zirconium" in: *Ullmann,* 6th ed., Electronic Release 1999.

[344] O. Helmboldt et al.: "Aluminum Compounds, Inorganic" in: *Ullmann,* 6th ed., Electronic Release 1999.

[345] H. Stark: "Iron Chlorides" in: *Ullmann,* 6th ed., Electronic Release 1999.

[346] J. S. Sconce: *Chlorine—Its Manufacture, Properties and Uses,* Reinhold Publ. Corp., New York 1962, p. 805.

[347] W. Zulehner: "Silicon" in: *Ullmann,* 5th ed., **23,** 721–748.

[348] The Chlorine Institute, "Assessment of the Economic Benefits of Chlor-Alkali Chemicals to the United States and Canada", Charles River Associates, Boston, April 1993.

[349] W. Schneider: "Phosgene" in: *Ullmann,* 5th ed., **19,** 411–420, and 6th ed., Electronic Release 1999.

[350] G. Zelesny, Brochure "Prochlor", University of Cologne, 1999, p. 19.

[351] D. Dietrich, K. Uhlig: "Polyurethanes" in: *Ullmann,* 5th ed., **21,** 665–716.

[352] V. Serini: "Polycarbonates" in: *Ullmann,* 5th ed., **21,** 207–215.

[353] M. Rossberg, W. Lendle, A. Tögel: "Chloromethanes" in: *Ullmann,* 5th ed., **6,** 235–256, and 6th ed., Electronic Release 1999.

[354] G. Siegemund, W. Schwertfeger, "Fluorine Compounds, Organic" in: *Ullmann,* 5th ed., **6,** 247.

[355] European Centre for Ecotoxicology and Toxicology of Chemicals (ECETOC), Report no. 32, 1988 and 34, 1989 "Methylene chloride".

[356] Euro Chlor Fact Sheet "Chlorinated Solvents" Brussels, 1995.

[357] M. J. Molina: "The role of chlorine in stratospheric chemistry", *Pure & Appl. Chem.* **68** (1996) no. 9, 1749–1756.

[358] D. Carlson, W. Schmiegel: "Fluoropolymers, Organic" in: *Ullmann,* 5th ed., **11,** 393–429.

[359] Verband der Chemischen Industrie, Positionspapier "FCKW-Ausstieg vollzogen", Frankfurt/ Main, 1998.

[360] E. Dreher: "Chloroethanes" in: *Ullmann,* 5th ed., **6,** 257–282, and 6th ed., Electronic Release 1999.

[361] E. Dreher: "Chloroethylenes" in: *Ullmann,* 5th ed., **6,** 283–306, and 6th ed., Electronic Release 1999.

[362] M. Allsopp, G. Vianello: "Poly(vinyl Chloride)" in: *Ullmann,* 5th ed., **21,** 717–742, and 6th ed., Electronic Release 1999.

[363] G. Pyke, Euro Chlor Proceedings of the 3rd Global Chlor-Alkali Symposium Monte Carlo 1992, p. 208.

[364] Y. Vandendael: "Poly(vinyliden Chloride)" in: *Ullmann,* 5th ed., **22,** 17–29.

[365] ECETOC Report no. 60, "Trichloroethylene", 1994.

[366] ECETOC Report No. 37, "Tetrachloroethylene", 1990.

[367] G. Koenig, E. Lohmar: "Chloroacetic Acids" in: *Ullmann,* 6th ed., Electronic Release 1999.

[368] R. Jira, E. Kopp, B. McKusick: "Chloroacetaldehydes" in: *Ullmann,* 5th ed., **6,** 527–536 and 6th ed., Electronic Release 1999.

[369] E. Langer: "Chloropropanes" in: *Ullmann,* 5th ed., **6,** 307–311, and 6th ed., Electronic Release 1999.

[370] L. Krähling, J. Krey, G. Jakobson, L. Miksche: "Allyl Compounds" in: *Ullmann,* 6th ed., Electronic Release 1999.

[371] G. Liu et al.: "Chlorohydrins" in: *Ullmann,* 5th ed., **6,** 565–576, and 6th ed., Electronic Release 1999.

[372] K. T. Rowbottom: "Epoxides" in: *Ullmann,* 5th ed., **9,** 539–540, and 6th ed., Electronic Release 1999.

[373] J. W. Muskopf, S. B. McCollister: "Epoxy Resins" in: *Ullmann,* 5th ed., **9,** 547–563, and 6th ed., Electronic Release 1999.

[374] G. Jakobson, F. Kathagen, M. Klatt, U. Steinberner: "Glycerol" in: *Ullmann,* 5th ed., **12,** 477–489,
and 6th ed., Electronic Release 1999.

[375] D. Kahlich, U. Wiechern, J. Lindner: "Propylene Oxide" in: *Ullmann,* 5th ed., **22,** 239–260, and 6th ed., Electronic Release 1999.

[376] H. Rassaerts: "Chlorobutanes" in: *Ullmann,* 5th ed., **6,** 312–313.

[377] P. Kleinschmidt, H. Rassaerts: "Chlorobutenes" in: *Ullmann,* 5th ed., **6,** 314–322, and 6th ed., Electronic Release 1999.

[378] H. Strack, R. Cook: "Chlorinated Paraffins" in: *Ullmann,* 5th ed., **6,** 323, and 6th ed., Electronic Release 1999.

[379] Euro Chlor, "Chlorinated Paraffins — Toxicological and Environmental Questions & Answers", Brussels, 1996.

[380] U. Beck: "Nucleus-Chlorinated Aromatic Hydrocarbons" in: *Ullmann,* 5th ed., **6,** 330–354, and 6th ed., Electronic Release 1999.

[381] Euro Chlor, Risk Assessment "Monochlorobenzene", Brussels, 1999.

[382] F. Muller, L. Caillard: "Chlorophenols" in: *Ullmann,* 5th ed., **7,** 1–8, and 6th ed., Electronic Release 1999.

[383] NIOSH, Washigton D.C., US Government Printing Office, 1977, pp. 40-49.

[384] EC Council Directive 76/769/EEC of July 27, 1976.

[385] K. Lipper: "Side-Chain-Chlorinated Aromatic Hydrocarbons" in: *Ullmann,* 5th ed., **6,** 355–367, and 6th ed., Electronic Release 1999.

[386] R. Nolte, Euro Chlor 4th Technical Seminar in Leipzig, February 1997, pp. 70–89.

[387] H. W. Sidebottom, J. A. Franklin, *Pure & Appl. Chem.* **68** (1996) no. 9, 1757–1769.

[388] Euro Chlor Brochure "20 Questions and Answers about Dioxins", Brussels, August 1998.

[389] K. Ballschmiter, Ch. Rappe, A. Hanberg, *Pure & Appl. Chem.* **68** (1996) no. 9, 1771–1800.

[390] Verband der Chemischen Industrie, Positionspapier "Dioxine", Frankfurt, 1998.

[391] World Chlorine Coucil: Brochure "Dioxins and Furans in the Chemical Industry", January 1998.

[392] L. Krämer et al.: Legal Aspects, Chap. 5 "The Seveso Directive: An Example of the Interface between Community Law and National Implementation Rules" in: *Ullmann,* 6th ed., Electronic Release 1999.

[393] Euro Chlor "Fact Sheets about Chlorine – Dioxins", Brussels, 1995.

[394] http://irpc.unep.ch/pops/CEG-1/WG52aIIE.html

[395] ICCA - International Council of Chemical Associations: Brochure "Briefing Notes on Persistent Organic Pollutants", 1998.

[396] Euro Chlor "Chlorine Industry Review 1995–96", p. 15.

[397] D. Decuyper, *Euro Chlor Proceedings of the 4th Technical Seminar in Leipzig,* February 1998, pp. 101–116.

[398] Noren et al., *Environmental Health Perspectives* **104** (1996) 766.

[399] N. Takeuchi, *Soda to Enso* **39** (1988) no. 461, 277–290.

[400] Bertrams, Concentration Plants for NaOH-, KOH-, Na_2S- and $CaCl_2$-liquors, Muttenz, Switzerland, 1979.

[401] Sulzer-Escher-Wyss, US 4 927 494, 1990 (R. Winkler et al.).

[402] Oxy-Occidental Chem. Corp., Caustic Potash Handbook, Irving 1987.

[403] Oil Paint Drugs, Chemical Marketing Reporter, 28th May, 1990.

Index

JOHN RYLANDS
UNIVERSITY
LIBRARY OF
MANCHESTER